The Third Dimension
in Organic Chemistry

The Third Dimension in Organic Chemistry

Alan Bassindale

Department of Chemistry
The Open University

John Wiley and Sons
Chichester . New York . Brisbane . Toronto . Singapore

Published by John Wiley & Sons Ltd in cooperation with
Open University Press

British Library Cataloguing in Publication Data:

Bassindale, Alan
    The third dimension in organic chemistry.
    1. Stereochemistry   2. Chemistry, Physical organic
    I. Title
547.1'223   QD481

ISBN 0 471 90189 X

Photoset by Enset Ltd,
Midsomer Norton, Bath, Avon.
Printed in Great Britain by
M. & A. Thomson Litho Limited,
East Kilbride, Scotland.

Contents

Part I   Stereochemical Principles

Chapter 1   Observed molecular geometries in carbon compounds

Chapter 2   Bonding in organic compounds

Part II The Application of Stereochemical Principles

Chapter 12 The stereochemistry of elimination and addition reactions

Chapter 13 Pericyclic reactions

Chapter 14 Asymmetric reactions—stereodifferentiation reactions

Preface

Organic chemistry is a massive subject that is founded on four million or more known compounds, and countless chemical reactions.

Clearly, if order is to be discerned in organic chemistry, then we must look for patterns and generalizations from which hypotheses can be formulated and theories eventually built.

The title of this book is both a description of *stereochemistry*—or chemistry in space—and a metaphor for the way in which stereochemistry dominates organic chemistry. Molecular shapes and their relationship to reactivity and physical properties provide a single coherent thread that winds its way round and through the body of organic chemistry.

I have divided the book into three sections. The first comprising Chapters 1–5, provides the fundamental definitions and framework necessary for an understanding of molecular shapes, and the relationships between molecules (isomerism). There is a brief chapter on chemical bonding, taken at a very simplified level, and aimed only at students with a very slight background in bonding theory. Sufficient information is given to enable the main stereochemical features of molecules to be understood. Dynamic processes within molecules are very important in organic chemistry and an attempt has been made to highlight this in the first section.

The second section (Chapters 6–9) builds on the first by exploring in some detail the stereochemical behaviour of open-chain compounds, cyclic compounds, chiral compounds and some giant molecules.

Many traditional texts end with the study of giant molecules but this denies the reader a look at chemical reactions from a stereochemical view. The final third of the book, section three (Chapters 10–14), is concerned with the stereochemistry of some common reactions, including substitution, elimination, addition and pericyclic reactions. Each type of reaction is introduced and defined before it is studied for its stereochemistry.

The final chapter attempts to summarize some of the most exciting chemistry that can be conceived—that is the designed synthesis of molecules with a handedness or chirality.

Throughout the writing of this work the problem has been what to leave out, rather than what to include. Eventually I settled for the inevitable compromise between coverage and depth.

The book is designed to be used in one of two ways. It can be the basis for a one- or two-term course in stereochemistry, or it should be possible for students to use the book in parallel with other courses, on a self-paced basis. There are extensive summaries giving the main points of each section and worked answers to the problems are also given. Model use is essential for understanding stereochemistry and the book has been written with that in mind.

Acknowledgements

I would like to thank my colleagues at the Open University Chemistry Department for their encouragement and sympathy during the protracted period of writing this work. Dr. P. G. Taylor was particularly helpful with Chapter 10.

Lorna Cattell typed much of the first drafts, and following my remove to my present office in the Wimpey II hut, Anne Earls and Shirley Foster transformed the much-altered early work into its final form. Without their willing help I could not have completed the project.

Much of the book was written at home, and I thank Anna and our children for bearing with me, and the mountains of books and papers that threatened to take over several rooms.

Finally I thank Dr. Howard Jones and the publishers for their care in production.

As ever, errors are the sole responsibility of the author.

PART I

Stereochemical Principles

CHAPTER 1

Observed molecular geometries in carbon compounds

1.1 Introduction

Text books frequently start a study of organic chemistry with a chapter on the theory of bonding in compounds, and then use this to rationalize molecular geometry. This is a rather curious approach as bonding theories were developed in response to our knowledge of molecular geometry. We have chosen a different approach. In this chapter we present you with the actual geometries of some simple compounds and follow this in the next chapter with an account of a simple theory that helps to rationalize the observed geometry of molecules.

Our starting point is the assumption that molecules can be represented by geometric structures. There is some academic dispute about this currently but the organic chemist has no such worries—we assume that molecules have size and shape!

Two of the parameters defining molecular geometry in covalent compounds are *bond length* and *bond angle*.

The bond length is defined as the distance between two atomic nuclei, joined together by one or more covalent bonds. Molecules are constantly in motion, and one type of motion is vibrational, with the result that bond lengths are not constant. Values of bond lengths that are measured and quoted are mean, or equilibrium bond lengths.

The internuclear bond angle ABC is defined as the angle subtended by the three nuclei bonded together in the unit A—B—C. Again, bond angles are subject to variation owing to intramolecular deformations and averaged values are measured and quoted.

1.2 Idealized geometries for organic compounds

It is a general principle that when there are a number of atoms or groups, all bonded to the same central atom, those groups or atoms tend to arrange themselves so that they are as far apart from each other as possible within the constraints of the given bond lengths. This principle allows us to predict the approximate geometry about an atom. The most common examples for organic chemistry are now discussed.

For carbon atoms the geometry is usually very closely related to the *coordination number* at that carbon. The coordination number is the number of atoms directly bonded to the carbon atom in question.

The most common coordination numbers in organic chemistry are 4, 3, 2 and 1. Four coordination at carbon is generally defined by *tetrahedral geometry*. A regular tetrahedron is shown in Figure 1.1(a). For the compound Cabcd the carbon atom is considered to be at the centre of the tetrahedron with the groups abcd at the vertices. This is the arrangement by which all four groups are separated by the greatest possible distance. The assumption of regular tetrahedral geometry implies that all bond angles are exactly 109° 27′ and that all bond lengths are identical. This is an approximation that we shall examine in later sections of this chapter. In fact, small deviations from perfectly tetrahedral geometry are the rule rather than the exception; it is however a useful starting point for establishing molecular geometries. Similar arguments hold for 3 and 2 coordination.

For ease of representation, tetrahedra are not usually drawn and the more usual way of depicting tetrahedral geometry is shown in Figure 1.1(b). This type of representation is extremely useful and common and is called the '*flying wedge*' notation. Chemical bonds are represented by lines joining two points and three dimensions can be implied in the figures. A single line —— joins two chemically bonded atoms, both of which lie in the paper plane, or a plane parallel to the paper plane. A wedge joins two atoms that are not both in the paper plane. The atom at the thick end of the wedge is closer to the observer than the atom at the thin end, so a wedge represents a bond coming out of the paper plane. A dashed line – – – represents a bond going into the plane of the paper. It is usually apparent which atom is farthest away from the observer and in many cases it is unambiguous as a dashed line is associated with a wedge—as in Figure 1.1(b).

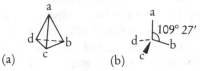

Figure 1.1   The tetrahedral geometry of a molecule Cabcd with 109° 27′ bond angles.

In ambiguous cases a hatched wedge ||||⋯ may be used. The atom farthest away from the observer is at the thin end of the wedge. Carbon atoms at the intersection of lines or wedges are frequently omitted for clarity. *Heteroatoms* (atoms other than carbon) such as oxygen and nitrogen should not be omitted. The flying wedge notation will be used extensively in the rest of the text to specify molecular geometry in three dimensions.

Three-coordination at carbon is defined by *trigonal geometry*. For a molecule or molecular fragment Cabc all four atoms are coplaner with all three bond angles being 120° in the idealized form as shown in Figure 1.2. The trigonal, planar form for Cabc again allows a, b and c to be as widely separated as possible.

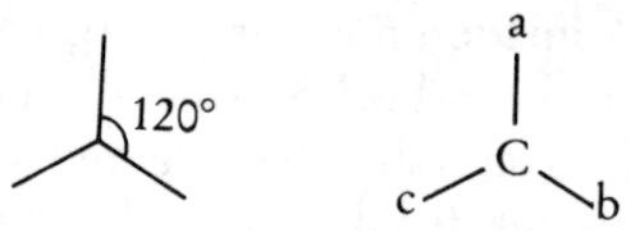

Figure 1.2 Trigonal geometry for Cabc with 120° bond angles.

Two-coordination at carbon is defined by digonal geometry in which a molecule or molecular fragment aCb has all three atoms colinear with bond angles of 180° as shown in Figure 1.3.

Figure 1.3 Digonal geometry for Cab with 180° bond angles.

One-coordination is trivial as there is only one way of joining two atoms Cb with a fixed bond length. One coordination at carbon is rare in organic chemistry, being limited to carbon monoxide CO and ions such as cyanide, CN^-.

The geometry at heteroatoms is not quite so straightforward as for carbon, owing to the presence of lone pairs or non-bonding pairs of electrons. The number of lone pairs is easily determined by application of the Lewis method of electron counting. The idealized geometries may still be used for atoms with lone pairs if each lone pair is considered to occupy one coordination site around the central atom. For example, two-coordinate oxygen is not digonal but tetrahedron-based with approximately 109° bond angles. A molecule or fragment Oab has the atoms a and b, and two lone pairs surrounding the central oxygen atom. Therefore for the estimation of bond angles the oxygen in Oab may be considered to be four-coordinate and, to a first approximation, tetrahedral. Sulphur may be considered similarly.

Nitrogen, in organic compounds, may have a coordination number of 1 to 4. Four-coordinate nitrogen, associated with amine salts, is tetrahedral.

Three-coordinate nitrogen in the species Nabc is also tetrahedron-based as there are three groups or atoms and a lone pair surrounding the nitrogen atom. The lone pair of electrons cannot easily be observed and it is more usual to describe the geometry at three-coordinate nitrogen as *pyramidal* although the bond angles in the ideal case are still 109° 27′ as shown in Figure 1.4. The Lewis approach shows that in two-coordinate nitrogen Nab there is one lone pair in addition to the two atoms a and b, so the idealized geometry is trigonal derived with an aNb bond angle of 120°.

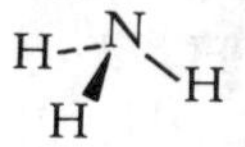

$$H{-}{-}\overset{N}{\underset{H}{\diagup}}{\diagdown}{H}$$

Figure 1.4   Pyramidal, tetrahedron-derived geometry for NH_3.

1.3 Molecular models

The idealized geometries described above are realistic approximations to actual geometries and are particularly useful in the building of molecular models. For the most effective use of this text you will need to have constant access to a molecular model kit and most of the model exercises are based on the *Orbit Molecular Model Kit*. The Orbit Kit consists of a number of plastic colour-coded centres to represent atoms. Each centre has spines symmetrically arranged around the surface with the appropriate number and distribution associated with coordination numbers 1–6. These atomic centres are equivalent to the idealized geometries with bond angles of 180°, 120°, 109° 27′, etc. Models of molecules are constructed by joining atomic centres together with plastic straws to represent chemical bonds. For most purposes it is sufficient to use straws of equal lengths (usually 3.5 or 5.0 cm) to represent bonds. If more accurate models are needed to represent particular molecules straws can be cut to scale and in certain cases atomic centres with different bond angles are available.

1.4 Actual geometries in organic compounds

The specific geometries of some organic compounds will now be examined to determine their deviation from ideality. You will see that in most cases deviations are relatively small, justifying the approximation to ideality. Where there are large deviations there is usually a sound reason, as for example with small ring formations.

The bond lengths for selected compounds are also introduced here. You should note that bond lengths are very small indeed falling in the range 75–200 pm where 1 pm (1 picometre) is 10^{-12} metre. The size of chemical bonds is hard to imagine but consider that Orbit molecular models are about 3×10^{10} bigger than actual molecules.

Table 1.1 gives a list of representative bond lengths for organic compounds. Note that bond lengths for similar bonds, whatever the total molecular structure, are similar. This observation can be used to estimate bond lengths in most compounds.

Table 1.1. Bond lengths in organic compounds

Bond type	Compound type	Bond length/pm
C—H	RCH_3	109.6±0.5
	C=C⟨H	108.3±0.5
	—C≡C—H	105.5±0.5
	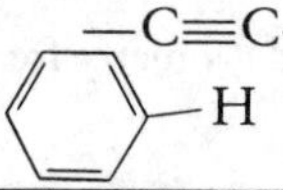 ⬡—H	108.4±0.6
C—F	all C—F compounds	135.3±2.6
C—Cl	all C—Cl compounds	170.1±6
C—Br	all C—Br compounds	186 ±7
C—I	all C—I compounds	210 ±10
C—C	alkanes	153.7±0.5
	C—C=	151.0±0.5
	C—C≡	145.9±0.5
	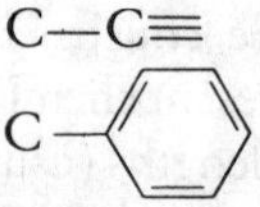 C—⬡	150.5± 0.5
C=C	alkenes C=C	133.5±0.5
	 C—C ⬡	139.4±0.5
C≡C	alkynes C≡C	120.2±0.5
C—N	Amines, C—N⟨ quaternary amines C—N⟨⁺	147.7±0.7
C=N	all C=N⟨	135 ±3
C≡N	C≡N	115.7±2

6

Table 1.1 (continued)

Bond type	Compound type	Bond length/pm
C—O	C—O alcohol ether	142.6±1
	⬡C—OH	136 ±1
C=O	aldehydes RCHO, ketones R_2CO acids RCOOH, esters RCOOR'	121.5±0.5 123.3±0.5
C—P	phosphines R_3P	184.1±0.5
C—S	thioethers R_2S	181.7±0.5

A consistent set of numbers called *covalent radii* have been calculated for atoms in covalent compounds. Some covalent radii are given in Table 1.2.

Table 1.2. Covalent radii of elements

Element	Coordination number of the bonded atom	Covalent radius/pm
C	4	77.0
	3	66.7
	2	60.3
N	3,4	70.0
	2	62.0
O	2	66.0
	1	54
H	1	33
F	1	64
Cl	1	99
Br	1	114
I	1	133

Covalent radii r_A are calculated so that a weighting is given to the contribution of an atom A to the bond length A—B. The atom B also has a covalent radius r_B, and $r_A + r_B$ gives the bond length AB. It is convenient that covalent radii are approximately independent of the nature of the other atoms bonded to the atom in question. For any bond A—B a close estimate of the bond length can be obtained from the sum of r_A and r_B.

Some specific compounds will now be examined. The geometry for methane CH_4 is shown in Figure 1.5. Methane, at equilibrium, is a perfectly tetrahedral molecule with bond angles of 109° 27′ and all four C—H bond lengths of 109 pm. It is a perfectly tetrahedral molecule because it is symmetrical; all four hydrogen atoms are indistinguishable. Only carbon atoms surrounded by four identical *ligands* (groups or atoms) can be expected to be perfectly tetrahedral.

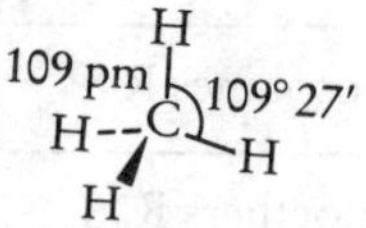

Figure 1.5   The tetrahedral geometry of methane.

Propane C_3H_8 does not have carbon atoms in which all four ligands are identical and the geometry about the central carbon atom is shown in Figure 1.6. The C—C—C bond angle of 112° is rather larger than the tetrahedral angle, and is typical for the angle between three adjacent carbon atoms in alkanes. Bond angles for similar molecular fragments are usually similar whatever the total molecular environment and some values are given in Table 1.3. The reasons for this and other deviations from ideality are discussed in the next Section.

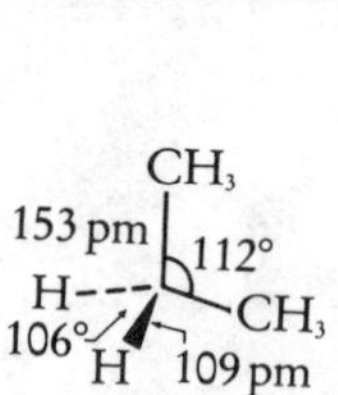

Figure 1.6   The distorted tetrahedral geometry of propane.

The geometry at heteroatoms such as oxygen and nitrogen that are expected to have tetrahedral geometry (bond angles of 109° 27′) is quite close to the ideal although deviations are frequently greater than at tetrahedral carbon. Some typical examples of geometry at two-coordinate oxygen are given in Figure 1.7.

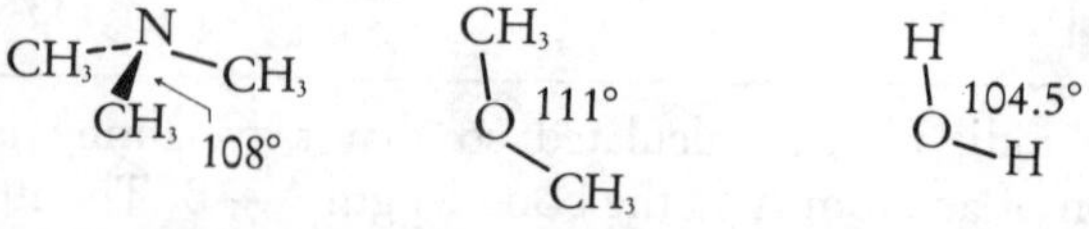

Figure 1.7   The geometry of trimethylamine, dimethyl ether and water.

Table 1.3. Bond angles in organic compounds

A—B—C	Approximate bond angle/°
C—C—H	107.8
C—C—C	112.6
H—C—H	105
C—C—C=	110.5
C—C—C≡	110
C—C=C	122 – 5
C=C⟨H	119
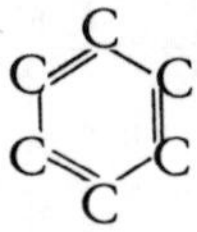 (benzene ring)	120
C—O—H	108.5
C—O—C	110±3
C—N—C	110
C∖C=O	121
H∖C=O	119
—C≡C—	180

The idealized geometry for alkenes RR'C=CR''R''' is trigonal with bond angles of 120° but ideality is rarely achieved. The parent alkene, ethene C_2H_4, has all six atoms in the same plane and the geometry shown in Figure 1.8. There is a similar geometry about the carbon atom in the planar molecule methanal, H_2CO, although the bond angles in carbonyl compounds (compounds containing C=O) is very sensitive to the ligands around the carbon atom, as shown in Figure 1.9.

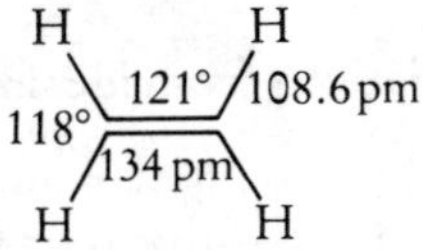

Figure 1.8 The planar geometry of ethene.

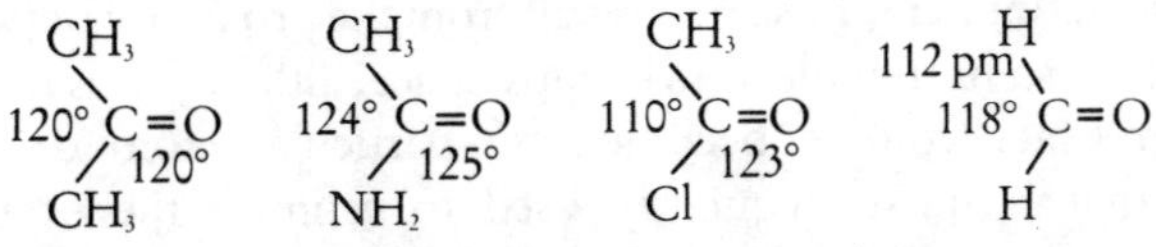

Figure 1.9 The geometry of some carbonyl compounds.

The bond angles at 2-coordinate carbon are usually 180° in agreement with the idealized geometry. Ethyne HC≡CH has 180° bond angles and a C—H bond length of 105.8 pm and a C—C bond length of 120.5 pm.

The few examples of actual molecular geometries given above illustrate that although ideality in geometry is rare, deviations are small. The example with the greatest deviation from ideality is cyclopropane, C_3H_6, as shown in Figure 1.10. In this example the internuclear bond angle is 60° as necessitated by the formation of a regular three-membered ring. The exceptional example of cyclopropane is referred to again in Chapter 2 where the nature of its bonding is discussed.

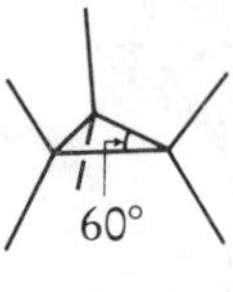

Figure 1.10   Cyclopropane geometry.

1.5 Summary of Sections 1.1–1.4

The following general points concerning molecular geometries are worth noting.

1. The most common idealized geometries in organic chemistry are; tetra-hedral, trigonal and digonal.
2. Molecular geometries deviate from ideality but not sufficiently to negate the usefulness of describing molecules in terms of their ideal geometries.
3. Bond angles for the same groups of atoms (e.g. C—C—C in alkanes), whatever the total structure of the molecule, are found to have similar values (Table 1.3).
4. Bond lengths for pairs of atoms can be estimated from the sums of the covalent radii r_A and r_B. Table 1.1 shows typical bond lengths and Table 1.2 calculated values of covalent radii.
5. Very small ring compounds (3 and 4 members) suffer the greatest deviations from ideality owing to geometrical necessity.

1.6 Rationalization of deviations from idealized bond angles

Bond angles are influenced by two types of interactions; *steric* and *electronic*. The main interactions are *geminal*—that is, interactions between two ligands bonded to the same carbon atom.

Steric interactions are those that result from the proximity of two, or more, large ligands. Atoms in molecules occupy space, although it is not possible to measure an exact volume occupied by particular atom or group. The parameter that is most frequently used to estimate the extent of steric interactions is the *van der Waals* radius of a ligand. In crystals and liquids, molecules are attracted to one another by *van der Waals forces* which are electrostatic in origin and may be through permanent *dipoles* in molecules, or rapid and random fluctuating dipoles arising from electron cloud fluctuations. When atoms are brought close together the attractive van der Waals forces are opposed by repulsive forces, as the electron clouds of the two atoms interpenetrate. The distance at which the repulsive and attractive forces are balanced determines the van der Waals radii. For example the van der Waals radius of an element X is half the distance separating two contiguous but non-bonded X atoms in a crystal. Van der Waals radii for some elements and groups are given in Table 1.4.

Table 1.4. van der Waals radii

Atom or group	van der Waals radius/pm
H	120
N	150
O	140
S	185
F	135
Cl	180
Br	195
I	215
CH_3	200

If two bulky groups attached to a tetrahedral carbon atom are separated by a distance approaching the sum of their van der Waals radii they tend to repel one another, and consequently the bond angle at the central carbon increases beyond 109° 28'. In general, when the bond angle between one pair of groups on a tetrahedral carbon atom is increased to a value above 109° 28' then the bond angle between the other pair of groups decreases below the tetrahedral angle. This observation is often called the *Thorpe–Ingold* effect, and is shown schematically in Figure 1.11. The geometry of propane was shown in Figure 1.5 and this is a simple example of Thorpe–Ingold effect. The methyl groups have a van der Waals radius of about 200 pm compared with that of 120 pm for hydrogen. The methyl groups interact so that the C—C—C angle increases to 112° and the H—C—H angle reduces to 106°.

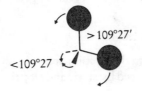

Figure 1.11   Schematic illustration of the Thorpe–Ingold effect.

11

Electronic interactions generally arise from the repulsion of similar charges. Electrons in adjacent covalent bonds repel with consequent effects on molecular geometry. In C—H bonds, which are very short, the electrons are relatively close to the carbon atom which means that in the absence of other effects there tends to be a widening of the C—H bonds above 109° 28' as a result of this repulsion. In bonds such as C—Cl the electrons in each bond are further from the carbon nucleus and therefore further from each other.

The electronic effect is observed in dichloromethane CH_2Cl_2 where the H—C—H bond angle is widened to 112° and the Cl—C—Cl angle is reduced to 108°.

It is not always easy to predict whether steric or electronic effect will dominate. Propane, again, provides a good example. In that case the steric effect of the methyl groups outweighs the electronic effect of the C—H bonds and the C—C—C bond is widened.

Lone pairs of electrons are relatively close to the nucleus and have a large electronic effect. The order of decreasing repulsion between pairs of electrons is lone pair–lone pair > lone pair–bond pair > bond pair–bond pair.

Consider a molecule such as water, H_2O, which was stated to be tetrahedron-based; that is the atomic framework is bent with an approximate bond angle of 109° 28'. As the lone pair–lone pair repulsion is greatest, the angle between lone pairs should increase—although such an angle cannot be measured—with a corresponding decrease in the H—O—H bond angle. The H—O—H bond angle is 104.5° in accordance with this simple idea. Similarly in ammonia, NH_3, the largest interaction is the lone pair–bond pair interaction implying that the H—N—H bond angles should be compressed below 109° 28'. This is what is observed and the bond angle in ammonia is 107.3°. A conflict between steric and electronic effects is observed in both ethers R—O—R and amines $R_{3-n}NH_n$. The C—N—C bond angle in trimethylamine $(CH_3)_3N$ is approximately 109° so that steric and electronic influences are about equally balanced. The steric effect outweighs the electronic effect in dimethyl ether, where the C—O—C bond angle is 111.5°.

The same general idea of steric and electronic influences can be applied to trigonal systems and this is illustrated in problem 5 at the end of the chapter.

Digonal systems, at least when unconstrained by rings, show no distortion from 180° bond angles. Steric effects cannot operate to change bond angles in this case as the ligands are already as remote as possible.

1.7 Summary of Section 1.6

The important points from this section are:

1. Bond angles are influenced by steric and electronic effects through geminal interactions.
2. Steric effects are those arising from the size of ligands. Two bulky groups

repel to widen the bond angle between them. This widening is usually accompanied by a decrease in the bond angle between the other pair of ligands. (Thorpe–Ingold effect.)
3. Electronic effects arise from van der Waals forces between the electron clouds of neighbouring ligands. When neighbouring electron clouds interpenetrate there is a repulsive interaction.
4. Electronic effects decrease in the order lone pair–lone pair > lone pair–bond pair > bond pair–bond pair.
5. Frequently electronic and steric effects are in opposition and it is not simple to decide which will dominate.

Problems and Exercises

1. In this book we shall concentrate heavily on the use of models to illustrate principles. The first exercise is to construct models of CH_4, $H_2C{=}CH_2$ and $HC{\equiv}CH$ so that you can see the dimensional implications of tetrahedral, trigonal and digonal geometries. The instructions for making single, double and triple bonds are contained in your model kit.
2. Without consulting the text fill in the blanks in the following Table:

Central atom	Coordination number	Name of geometry	Bond angles (idealized)
C	4		
C	3		
C	2		
N	4		
N	3		
N	2		
O	2		

3. Use a tetrahedral centre and four 3.5 cm straws to construct a model of methane. Use an octachedral centre (6 prongs) and four 3.5 cm straws to construct a *planar* CH_4 model. Measure the distances between all hydrogens in both models to show that tetrahedral geometry allows the greatest separation of ligands.
4. Make a model of CH_2Cl_2, using green centres at one end of the straws for chlorine atoms, and use it to help you draw flying wedge representations of that molecule from a number of viewpoints, for example:

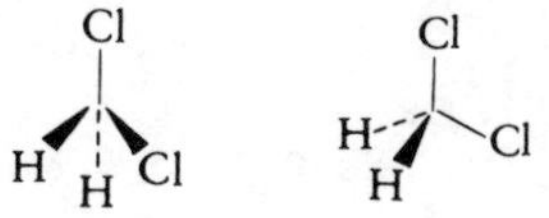

You may be surprised how many equivalent representations you can draw!

5. Rationalize the bond angles in the following compounds:

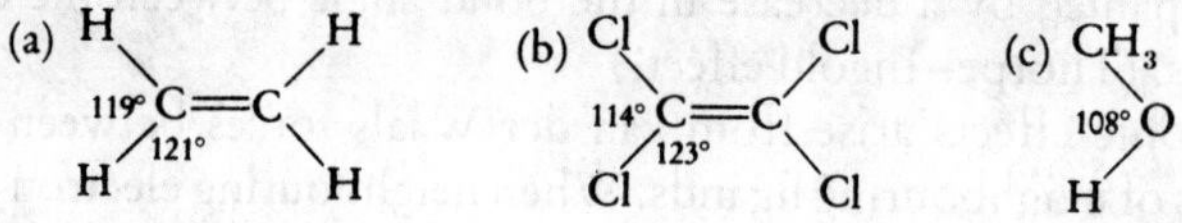

CHAPTER 2

Bonding in organic compounds

2.1 Introduction

Theories of chemical bonding have become extremely sophisticated and somewhat complex with the availability of high-speed computers. For most purposes simple, qualitative explanations of bonding suffice. In this chapter we concentrate on aspects of chemical bonding that are relevant to stereo-chemical problems. These fall into two categories; those dealing with the directional properties of chemical bonds and those concerned with dynamic aspects. In the first chapter we concentrated on molecules with a fixed geometry; if it was possible to photograph those individual molecules, a series of photographs over a period of time would all look very similar. Most molecules do not have fixed geometries in that sense as there is frequently rotation about chemical bonds. This aspect of molecular geometry is covered fully in Chapters 5 and 6 on conformations, but the explanation of how rotations about bonds can occur is found in this chapter.

2.2 Hydrogen-like atomic orbitals

Before chemical bonding is tackled it is useful to revise briefly atomic structure and electron distributions. The basis for this is usually taken to be the hydrogen atom for which exact solutions are available for the wave equations governing electron energies and distributions.

The atomic *nucleus* is made up of a number of positively charged *protons* and neutral *neutrons*. In an atom the number of negatively charged *electrons* surrounding the nucleus is equal to the number of protons in the nucleus, ensuring electrical neutrality. Protons and neutrons are adequately des-cribed as particles. Early models of atoms assumed that electrons were minute point charges describing fixed orbits around the nucleus. This model was

15

found to be inadequate and has been revised to allow for the wave nature of electrons. It is not possible to describe the exact position of an electron of given energy and the electron distribution around a nucleus is described in terms of a *probability function*. A particular three-dimensional region of space is defined in which there is an 80–95% probability of finding the electron at any given time. These regions of space are called the *atomic orbitals*.

Electron energies are *quantized*, the electrons in an atom must have an energy value selected from a given set. Intermediate energies are forbidden.

We may now turn to the *hydrogen atom* which has only one electron. The lowest energy state, or *ground state* of the hydrogen atom has the electron in a spherically symmetric atomic orbital as shown in Figure 2.1. This orbital is labelled as $1s^1$. The first 1 refers to the *principal quantum number n* of the orbital, with the energy of the orbital increasing as n increases. Orbitals with $n = 1$ are described as being in the 'K shell' and the K shell contains only the $1s$ orbital. The s describes spherically symmetric orbitals and the superscript 1 denotes the occupancy of the orbital by one electron. *Any atomic orbital can contain a maximum of two electrons.* (Pauli exclusion principle.)

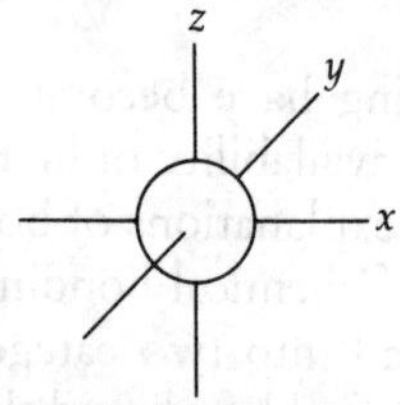

Figure 2.1   The spherically symmetrical $1s$ atomic orbital.

If energy is put into a ground state hydrogen atom then the electron can be *promoted* to higher energy states. The next highest orbital is the $2s$ orbital. This orbital is in the 'L shell' with $n = 2$ and is also spherically symmetric as designated by the label s.

Further promotion leads to occupancy of the $2p$ orbitals. There is one s orbital and three p orbitals all in the shell with $n = 2$. The three dumbell-shaped p orbitals are of equal energy and designated p_x, p_y and p_z, and are illustrated in Figure 2.2. These p orbitals are not spherically symmetric, but a cross-section is radially symmetric about the particular axis for which they are labelled. The two lobes of a p orbital are given opposite signs (+ or −) which refer to their *phase*. Similarly a $1s$ orbital can have a phase of either sign. The phase of an orbital does not describe energy or charge distribution, but is a mathematical description needed for bonding theory. The phases of orbitals may be described by analogy with a standing sine wave, peaks have one phase and troughs the opposite.

Promotion from the L shell to the M shell with principal quantum number 3 leads to the $3s$ and $3p$ orbitals and then to a set of orbitals called 'd orbitals' of which there are five of equal energy. These d orbitals do not figure greatly in organic chemistry and we shall not discuss them further.

16

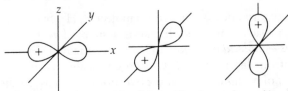

Figure 2.2   The p_x, p_y and p_z atomic orbitals. The signs refer to the orbital phases.

It is assumed that all atoms have hydrogen-like atomic orbitals and that the occupancy of these orbitals in atoms follows well-defined rules. The rule may be illustrated by reference to carbon which has six electrons. *The first rule is that the lowest energy orbitals are filled first.* The lowest energy orbital is the 1s orbital and this is filled with two electrons to give 1s². The Pauli exclusion principle tells us that *two electrons are allowed in one atomic orbital if they have opposing spins.* The property of spin of an electron is difficult to describe simply and for our purposes we simply state that an electron can have one of two orientations of spin usually designated ↑ and ↓. The occupancy of the 1s orbital is thus represented, 1s ⊞. The orbitals are now filled sequentially and the next highest energy orbital is the 2s orbital which is filled, 2s ⊞. The next rule needs to be applied to the remaining two electrons. A problem arises as there are three 2p orbitals of equal energy and only two electrons. *Hund's rule* states that where there are several orbitals of equal energy *(degenerate orbitals)* the *electrons are put first into separate orbitals with unpaired spins.* The basis of this rule is that as electrons are negatively charged they repel one another, and two electrons in one orbital exert a greater repulsive effect than electrons in separate orbitals. By making the spins unpaired the electrons cannot occupy the same orbital at a given time (Pauli principle). Hund's rule applied to carbon gives the *sp* occupancy as 2p ⊞. It is pointless to label the occupancy of these atomic orbitals as $2p_x{}^1 2p_y{}^1$, or any other combination as the coordinate system is imposed by us and not well-understood by the electrons!

The atomic orbital description of a ground state carbon atom is therefore 1s²2s²2p² or 1s ⊞ 2s ⊞ 2p ⊞

The rules stated above may be applied to any atom to determine its electron configuration. In the next section we shall consider chemical bonding and only the electrons in the outermost shell (the L shell for carbon) are of concern.

2.3 Chemical bonding

Atoms in molecules are held together by chemical bonds. Each discrete chemical bond in a stable molecule is considered to be made up of two electrons in a *molecular orbital* which is spread over two or more nuclei. The driving force for bond formation is that electron energies are lowered when they are *delocalized* from one atom to several. This principle can be illustrated

17

by looking at the formation of a hydrogen molecule, H_2, from two hydrogen atoms. The approach that we shall take is a simple *linear combination of atomic orbitals—molecular orbital* LCAO-MO method. Figure 2.3 illustrates the electron distribution in two ground state hydrogen atoms and shows the effect of bringing together the two atomic orbitals. As the two H $1s$ orbitals overlap they combine, either additively to give the elliptically shaped *bonding molecular orbital* (BMO), or with repulsion to give the two-lobed *antibonding molecular orbital* (ABMO). The first rule of LCAO theory is that *two atomic orbitals combine to give two molecular orbitals. In-phase combinations are bonding; out-of-phase combinations are anti-bonding.*

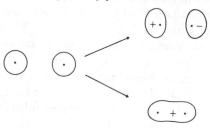

Figure 2.3   A simple LCAO picture of the hydrogen molecule. The lower, peanut-shaped orbital is a bonding orbital with electron density greatest between the nuclei (shown as points). The upper two-lobed orbital is antibonding with the lobes having opposite phases.

Now let us consider the energies of the two hydrogen molecular orbitals. As might be expected from the name, the bonding molecular orbital is of lower energy than the antibonding molecular orbital and the energy of the atomic orbitals is about mid-way between the two MOs. The extent of lowering of energy of the bonding MO is dependent on the efficiency of the *overlap* between the two atomic orbitals.

When it come to considering the occupancy of the moleculur orbitals in the hydrogen molecule, very similar rules to those for atomic orbitals are encountered. The lowest energy orbital is filled first with two spin-paired electrons. So, for the hydrogen molecule there are two electrons in the bonding molecular and none in the antibonding orbital. The overall energy of the hydrogen molecule is therefore lower than that of two isolated hydrogen atoms, accounting for the observation that hydrogen is found as H_2 not H. The two-electron bonding molecular orbital is usually referred to as a *covalent σ-bond.* Covalent means that the electrons are shared between two or more atoms and σ describes symmetry about the internuclear axis. The significance of this will soon become apparent.

It is also worth noting at this point that the familiar noble gas or inert gas rule is followed in the hydrogen molecule. This states that the most stable electron configurations for atoms in molecules is that of the nearest noble gas. Helium is nearest to hydrogen and has two electrons. Each hydrogen nucleus in the hydrogen molecule is surrounded by two electrons (although the two electrons are shared between the two hydrogen nuclei).

In the next Section the ideas of LCAO-MO are extended to carbon compounds.

2.4 Summary of the main points of Sections 2.2 and 2.3

1. Electron distributions in atoms and molecules are described in terms of probability functions called atomic or molecular orbitals. These define regions of space in which there is an 80–95% probability of finding a particular electron.
2. Electron energies are quantized, with energy, and shape of the probability functions being defined by the principal quantum number and symmetry label. The label 's' denotes a spherically symmetric orbital and 'p' a dumbell-shaped orbital.
3. Electron distributions in atoms and molecules are governed by the rules;
   (a) lowest energy orbitals are filled first;
   (b) two electrons with opposed spins are allowed in each atomic or molecular orbital;
   (c) for degenerate orbitals, electrons are put first into separate orbitals with opposed spins (Hund's rule).
4. Chemical bonds are formed when two or more atomic orbitals overlap to form molecular orbitals (LCAO-MO).
5. Bonding molecular orbitals are those with energy lower than the constituent atomic orbitals, *and are formed from in-phase combinations.*

2.5 Hybridization

In Chapter 1, methane, CH_4, was described as having four equivalent hydrogen atoms surrounding the carbon atom in a tetrahedral array. That is an experimental observation and the bonding theory that follows is a *rationalization* of the fact; not an explanation of why the tetrahedral cabon atom is found for four-coordinate carbon. The rationalization is useful, though, as it allows us to extend the ideas to other molecules and provides a simple picture of the observed stereochemistry of molecules.

The electron configuration of carbon is $1s^2 2s^2 2p^2$. We know that there are four C—H bonds in methane and each bond contains two electrons. Formally, the hydrogen atoms contribute one electron each, leaving the carbon atom to contribute four electrons. The noble gas rule is therefore fulfilled for carbon, in methane, with eight electrons in the L shell and two in the K shell, which is the neon configuration. Hydrogen again attains the helium configuration. For carbon and other elements of interest in organic chemistry the noble gas law can be restated that the outermost electron shell (*the valency shell*) should contain an octet of electrons in a stable molecule.

The problem now is to combine the carbon $2s^2$, $2p^2$ and the other $2p$ atomic orbital with the four hydrogen $1s^1$ orbitals so that the four C—H bonds are equivalent. The method used here is a mathematical abstraction called *hybridization*. Four equivalent bonding orbitals can be obtained from one s orbital and 3 p orbitals by adding them together *before* forming chemical bonds. The mathematical result of adding the valence shell orbitals of carbon is shown in Figure 2.4. Hybridization of the orbitals shown gives four equivalent orbitals called sp^3 *orbitals* each at an angle of 109° 27′ to the others. The total energy of the system is unchanged. It is now a simple matter to put one electron into each sp^3 orbital and bring up the four hydrogen atoms to produce methane. In this model, methane has four identical 2-electron C—H bonds (and of course four unoccupied antibonding orbitals). Each bond has σ-symmetry.

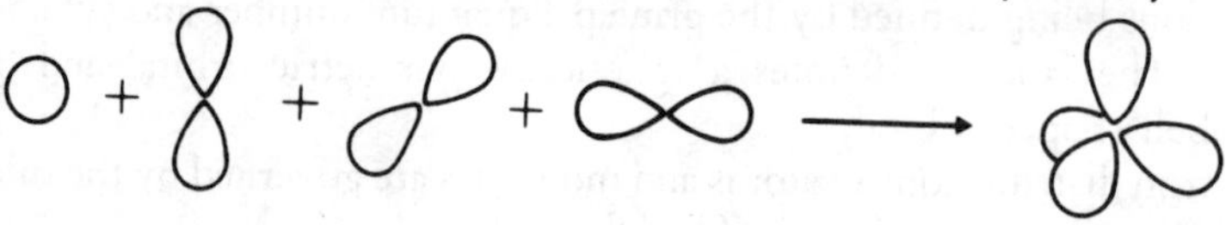

Figure 2.4　The construction of the hypothetical sp^3 orbitals from one s and the p_x, p_y and p_z atomic orbitals. Each lobe is at an angle of 109° 27′ to the others.

The sp^3 hybridized atom is the basis for the tetrahedral and tetrahedral fragment geometry described in Chapter 1.

Hybridization does not need to be complete, as in sp^3 hybridization, but can be adjusted according to our knowledge of the extent of multiple bonding in a molecule.

The sp^3 orbitals can be thought of as being built up in a sequential manner. If the s orbital and one p orbital are combined the result is sp hybridization in which two equivalent orbitals are produced that are separated by 180° as shown in Figure 2.5. It should not have escaped your notice that this is the geometry characteristic of two-coordinate carbon!

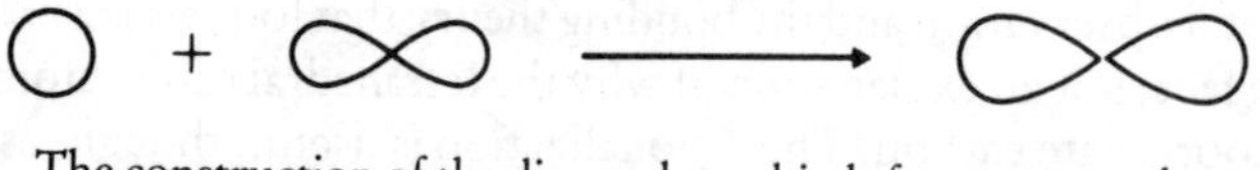

Figure 2.5　The construction of the diagonal sp orbitals from one s and one p orbital.

The next step is to add a p orbital to sp hybridization to give sp^2 hybridization as shown in Figure 2.6 in which three equivalent orbitals at 120° are formed. Three-coordinate carbon has bond angles of 120°.

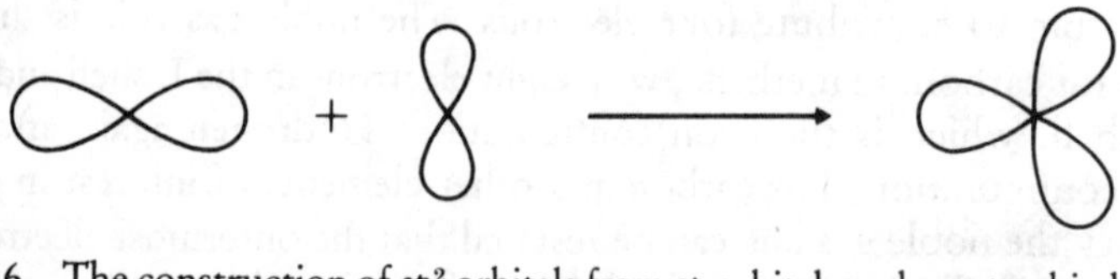

Figure 2.6　The construction of sp^2 orbitals from sp orbitals and one p orbital.

20

The final step is the addition of a p orbital to the sp^2 orbitals. The relevance of the hybridization model to molecular geometry in multiply bonded carbon compounds is shown in the next Section.

The application of sp^3 hybridization to amines and ethers is covered in the exercises at the end of the chapter.

2.6 π-bonds

Ethene, C_2H_4, is planar with bond angles of about 120° and is written $\begin{smallmatrix} H \\ H \end{smallmatrix} \!\! C{=}C \!\! \begin{smallmatrix} H \\ H \end{smallmatrix}$. The so-called *double bond* between the two carbons means that there are four electrons, in two pairs, shared between the two carbon atoms. The most convenient way of producing an MO picture for ethene, and one which satisfactorily accounts for its properties is shown in Figure 2.7. The starting point for this model is sp^2 hybridized carbon. One electron is put into each sp^2 orbital and the fourth is placed in the unhybridized p orbital. The two hydrogen atoms are then brought to two sp^2 orbitals to produce a CH_2 fragment. Now two CH_2 fragments are brought together with overlap of the two singly occupied sp^2 orbitals to produce a σ—C—C bond. The final step is to orientate the two singly occupied p orbitals so that they overlap in *sideways* manner. This is shown in more detail in Figure 2.8. Overlap of this type produces two new molecular orbitals; a bonding *π-type molecular orbital* and an antibonding π orbital usually labelled π^*.

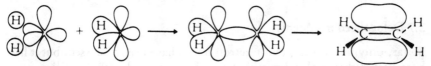

Figure 2.7   Sequential formation of a simple MO picture for ethene. Formation of a CH_2 unit from sp^2–pC and two hydrogen atoms is shown first. Then a C—C σ-bond is shown, Finally the π-bond MO only is shown.

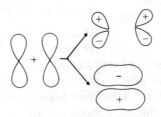

Figure 2.8   The overlap of two p orbitals in a sideways manner. The in-phase combination gives a π-bonding orbital and the out-of-phase combination gives a π^* anti-bonding orbital.

As there are only two electrons available, only the bonding orbital is filled. The label π for the bonding orbital is again a symmetry label indicating that the orbital is not spherically symmetric about the internuclear axis. The stereochemical consequences of this are profound and are mentioned in Section 2.11 and will occur throughout the book.

Ethyne is linear with bond angles of 180° and is written H—C≡C—H. The *triple bond* between the two carbon atoms implies that there are three, two-electron bonds between the carbon atoms. The geometry suggests that sp hybridization would be a convenient starting point and this is what is used. The hydrogen atom on each carbon is bonded to one sp orbital and the other used for a C—C σ-bond formation. There remain two p orbitals on each carbon atom which can form two π-bonds as shown in Figure 2.9. The combination of two π-bonds is spherically symmetric and the two carbon nuclei are enclosed in a cylindrical sheath of electrons.

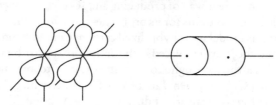

Figure 2.9 The formation of a triple bond from overlap of two sets of p orbitals. Bonding orbital only shown.

These principles of π-bond formation can readily be extended to such systems as C=N, C=O, C=S and C≡N.

2.7 Delocalized π-systems

So far, only two-centre, two-electron bonds have been discussed, but it has been stated that molecular orbitals can be associated with more than two nuclei. We shall now examine the chemical consequences of delocalization over more than two atoms. The first example is provided by the amides, of general formula $R—C{\overset{O}{\underset{NR^1R^2}{}}}$. There is a σ framework for amides consisting of R—C, C—O, C—N, N—R^1 and N—R^2 bonds. In addition there is a C—O π-bond and two lone pairs of electrons on the oxygen and one on the nitrogen atom. The oxygen lone pairs do not concern us here. The nitrogen lone pair, for amides in their simplest form, is located in an sp^3 orbital. It is observed in most amides that the major framework is planar, that is the N, C, O atoms and all those directly bonded to them are in the same plane. This is inconsistent with sp^3 hybridization at nitrogen but consistent with sp^2 hybridization. Figure 2.10 shows the orbital picture of addition of three equivalent, adjacent p orbitals. Three molecular orbitals are formed, one bonding, one non-bonding and one antibonding orbital. Let us now consider the application of this to amides. There are two electrons in the C—O π-bond and one lone pair on the nitrogen. When the new molecular orbitals are filled, starting with the lowest energy, the bonding and the non-bonding orbital are filled. This arrangement is of lower energy than an isolated π-bond and a lone pair. So the π-electrons in amides are delocalized over the three centres, C, N and O. In fact the electron density still remains larger in the C—O region than in the C—N region and it is said that amides have a *partial double bond* which is often written,

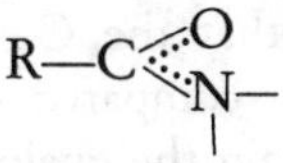

$$R—C\overset{O}{\underset{N—}{\diagdown}}$$

A more dramatic example of electron delocalization is provided by benzene, C_6H_6, whose structure has been shown to be planar, with a regular hexagon of carbon atoms in which all C—C bond lengths are identical, and half way between single and double bond length. If the carbon atoms are assumed to be sp^2 hybridized and are arranged in a regular hexagon, each carbon atom has one C—H σ-bond from one sp^2 orbital, and two C—C σ-bonds from the other two sp^2 orbitals. This leaves six p orbitals regularly arranged and with the correct orientation for π-overlap. There are six electrons available for bonding, which is equivalent to three π-bonds. The six p orbitals combine to give three bonding and three antibonding orbitals. The overall combination of the three occupied bonding orbitals gives an electron distribution rather like a ring doughnut above and below the ring. As there are three pairs of π-electrons to share between six bonds it is easy to see how the C—C bond length in benzene is intermediate between a single and a double bond. Benzene is frequently drawn, ⬡ to emphasize the electron delocalization. Benzene is an example of a set of cyclic compounds termed *aromatic* compounds, in which the delocalization has a stabilizing effect. Aromatic systems have *$4n+2$ π-electrons* in a cyclic array. Cyclic systems with *$4n$ electrons* are profoundly destabilized and are often said to be *anti-aromatic*. The reasons for this are beyond the scope of this book but some of the implications are discussed in Chapter 13.

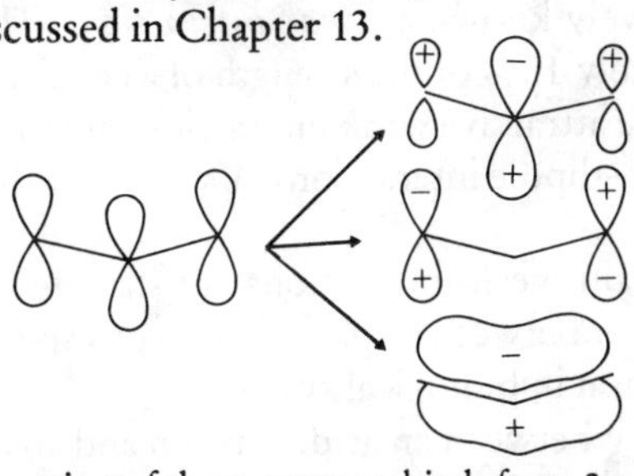

Figure 2.10   The formation of three-centre orbitals from 3 p orbitals.

2.8 Bonding in small ring compounds

In Chapter 1 the problem of the cyclopropane bond angles were raised. The tetrahedral bond angle appropriate to sp^3 hybridization and four coordination is 109° 27'. In cyclopropane the carbon—carbon bond angles cannot be other than 60°, and the compound is said to exhibit *ring strain*. Although the

23

internuclear bond angles are 60° in cyclopropane a molecular orbital model has been derived in which the *orbital bond angles* are about 109° 27' and sp^3 hybridization is retained. This model is illustrated in Figure 2.11. The overlap between the sp^3 orbitals in the arrangement shown is not as complete as in, say, straight chain alkanes, where the electron density is greatest along the internuclear axis. The result of this is that the C—C bond energies in cyclopropane are reduced by about 10% compared with those in acyclic alkanes. In accordance with this ring strain the cyclopropane ring is very susceptible to opening in reactions of the type,

$$\begin{array}{c} CH_2 \\ \diagup \quad \diagdown \\ CH_2 \text{—} CH_2 \end{array} + Cl_2 \rightarrow ClCH_2CH_2CH_2\,Cl$$

in which the ring strain is removed.

Cyclobutane has internuclear bond angles of 90° and is therefore strained. A similar bonding picture for this molecule can be drawn. Overlap in this case is greater than in cyclopropane and the bond energies are decreased by about 7–8% relative to acyclic alkanes. Ring strain in larger acyclic systems is quite small owing to ring puckering which is explained in Chapter 7.

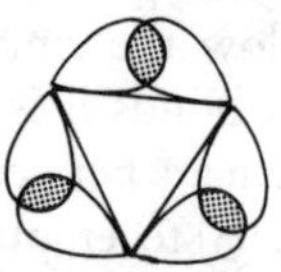

Figure 2.11   An orbital picture of the bonding in cyclopropane.

2.9 Weak interactions

Molecular geometry can be profoundly affected by intra and intermolecular interactions collectively known as *weak interactions*. These interactions may be up to approximately 10% of the strength of a covalent bond.

The weakest of the attractive weak interactions are the van der Waals forces which are the dipole–dipole interactions. Van der Waals forces are very short range in effect.

The other two main weak interactions are stronger and of longer range. *Electrostatic attraction* between oppositely charged species is non-directional and especially common in biological systems.

Hydrogen-bonding between an acidic proton and an electron donor group is both directional and long range. The bond, usually written A—H- - -B— is common for alcohol, amine, acid, and amide hydrogen atoms interacting with ethers, amines and fluorides. The direction can be estimated by assuming that the bond arises from an interaction with a lone pair on B with the hydrogen atom, along the line of the A—H bond. Hydrogen bonds (the H—B distance) are typically about 300 pm in length, and 10–30 kJ mol⁻¹ in energy.

Intramolecular hydrogen bonds can assist cyclization of small molecules and intermolecular hydrogen bonds encourage the association of two or more molecules. The fixed direction of hydrogen bonds have profound implications for the three-dimensional structures of molecules.

2.10 Summary of the main points of Sections 2.4–2.9

1. Stable carbon compounds normally follow the octet rule so that there are eight electrons in the valency shell.
2. Four equivalent, tetrahedrally disposed orbitals for an atom can be created mathematically by adding together 1 s orbital and 3 p orbitals in what is called sp^3 hybridization. This model is convenient for all four-coordinate compounds commonly found in organic chemistry.
3. sp^2 hybridization results in the formation of three trigonally disposed MOs and an unhybridized p orbital in a plane perpendicular to the trigonal plane.
4. sp hybridization results in two linearly disposed MOs and two un-hybridized p orbitals in planes perpendicular to the linear axis.
5. π-bonds are formed by sideways overlap of two p orbitals. π-bonds are not spherically symmetric about the internuclear axis.
6. Amides have three-centre, delocalized π-bonds formed by sideways overlap of three p orbitals.
7. Weak interactions, in increasing order of strength are, van der Waals forces, electrostatic interactions and hydrogen bonds.
8. Hydrogen bonds A—H– – –B, form between acidic hydrogens and electronegative atoms with lone pairs of electrons. They are directional and about 300 pm in length.

2.11 Rotations about bonds

This Section is of fundamental importance in setting the scene for many of the stereochemical properties of molecules discussed in later chapters. Each bond type has been brought together here to highlight the differences in rotational ability.

Single, or σ-bonds, have cylindrical symmetry about the internuclear axis. The consequence of this, in the case of, say, ethane CH_3CH_3 is that the extent of overlap of atomic of hybrid orbitals is unaffected by rotation about the carbon–carbon bond. Ethane can be thought of in terms of two linked CH_3 propellors, with each CH_3 rotating rapidly. This does not mean that all arrangements of ethane are of equal energy as there are other interactions to consider (Chapter 6). Nevertheless the energy difference between different arrangements is so small that there is generally very rapid rotation about any simple, σ-bond that is not constrained by rings or other geometrical constraints.

In compounds such as ethene there is a significant resistance to rotation about the double bond. The origin of this resistance is easily understood by looking at the extent of overlap of the two p orbitals as they are rotated relative to one another. Figure 2.12 illustrates the effect of rotation on overlap and shows that when the two p orbitals are at 90° there is essentially no overlap. For rotation to occur about a π-bond, sufficient energy is needed to compensate for the entire energy decrease on bond formation. As a typical π-bond has an energy of about 250 kJ mol^{-1} this is a formidable requirement, and not surprisingly rotation about most double bonds is not easily achieved.

Figure 2.12   The variation of π-bond overlap with rotation about the C—C bond.

The very well-defined spatial requirements of π-bonds are reflected in *Bredt's rule* which states that, in small bridged ring systems a double bond cannot exist in the bridgehead position. Figure 2.13 shows two bridged hydrocarbons with double bonds at the bridgehead. The small ring compound (bicyclo(2,2,1)hept-1-ene, Figure 2.13(a)) is incapable of existence whereas the larger ring compound (bicyclo(5,2,1)dec-1-ene, Figure 2.13(b)) has been synthesized. The basis for Bredt's rule is best verified by the use of models and is demonstrated in Problem 3 at the end of the chapter.

Figure 2.13   (a) Bicyclo(2,2,1)hept-1-ene which cannot be isolated. (b) Bicyclo (5,2,1)dec-1-ene which has been isolated.

Amides, and other compounds with partial double bonds provide particularly interesting examples for study of bond rotations. In the case of amides the C—N bond is intermediate in character between a σ-bond and a double bond (1 σ +1 π bond). The result of this is that rotation about the C—N bond is *restricted* and usually occurs so that the rates and energies of this rotation may be easily measured at readily accessible temperatures. There are also distinct chemical differences between amide nitrogen atoms and amine nitrogen atoms arising from this delocalization. There will be more on amide bond rotations in Chapters 5 and 6.

Finally, *alkynes*, with the two π-bonds have radial symmetry about the internuclear axis and therefore show very low barriers to rotation about that axis.

This short section has highlighted the *dynamic* aspects of molecular shape. It is of fundamental importance to recognize that although we may examine molecular structure by a number of physical methods the result of that measurement may be a function of the time-scale of that measurement. A useful analogy is with a camera operating at different shutter speeds. If a photograph was taken of a runner using a very, very short expoure time (say 1/5000 s) the resulting picture would essentially 'freeze' the motion and we could examine the arrangement of the runner's limbs at a given moment. If another photograph was taken one second later at the same speed the picture would be of the same runner but with a different arrangement of arms and legs. Now, if a photograph of the same runner was taken at an exposure of one second the photograph would give an *averaged* picture of the runner's constantly changing shape. Similarly with molecules, if we know the time-scale of measurement we can estimate the rate at which the molecule is changing shape, and we may be able to deduce the molecular geometry at a given time.

2.12 Summary of Section 2.11

1. There is easy rotation about σ-bonds, as they are symmetric about the internuclear axis.
2. There is a considerable resistance to rotation about π-bonds as the extent of overlap is reduced to almost zero when the two p orbitals are at right angles.
3. Bredt's rule states that small-ring compounds with a double bond at the bridgehead cannot be made.
4. Rotation about the C—N bond in amides is more difficult than rotation about a σ-bond but not so difficult as rotation about a normal π-bond.
5. There is easy rotation about the C—C axis in alkynes as two π-bonds at right angles together are symmetric about the internuclear axis.

Problems and exercises

1. Make models of ethane and ethene and observe the way that rotation about the central C—C bond in each case mirrors the description of ease of rotation in Section 2.8.
2. Draw simple MO pictures for ammonia and water and show how sp^3 hybridization is most appropriate for the observed geometry of these compounds. What type of MOs are the lone pairs occupying?
3. To help you understand Bredt's rule attempt to make molecular models of the compounds shown in Figures 2.13 (a) and (b). What is the problem in the construction of 2.13 (a)?

4. The compound ![structure] does not show the normal characteristics of amide π-delocalization. Make a model of the compound and explain why the nitrogen is more amine-like than amide-like.

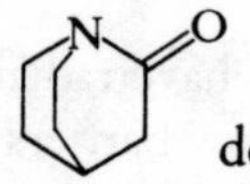

CHAPTER 3

Symmetry in organic compounds

3.1 Introduction

Any molecule or object can be classified in terms of its mathematical symmetry properties. Much of science is concerned with classification—the search for common properties or behaviour—so that some order can be perceived in the mass of known facts about the physical world. So, the labelling of molecules according to their symmetry properties enables us to understand and predict stereochemical properties and behaviour of molecules. There are two parts to this chapter. The first is concerned with *symmetry elements* within molecules. A symmetry element is a label given to a molecule that transforms in a given way when subjected to a *symmetry operation*. A symmetry operation is a way of interchanging geometrically equivalent parts of a molecule. There are four basic symmetry operations that will be studied. The importance of these symmetry operations and the corresponding symmetry elements is that possession of certain symmetry elements precludes one very important kind of stereochemical behaviour. Each of the symmetry elements will be studied and explained by the use of examples.

The second part of the chapter deals with the classification of molecular structure in terms of *point groups*. A molecule that possesses a particular ensemble of symmetry elements is given a single label called a point group that describes its overall symmetry. The point group classification finds significant usage in all branches of chemistry. In the remainder of the book molecules will occasionally be referred to by their point groups but the rather complex use of point groups and *group theory* is beyond the scope of this book. Point groups are therefore included as a preparation for further study or for interest.

3.2 Symmetry elements

3.2.1 Axis of symmetry C_n

If an imaginary line (*axis*) can be drawn through a molecule so that rotation by $360°/n$ gives a molecule indistinguishable from the original, that molecule is said to have a *rotation axis* C_n of order n.

The water molecule has a *two-fold* axis of rotation as shown in Figure 3.1(a). The C_2 axis is shown on the figure. The C_2 operation consists of rotating the molecule through 180° and comparing the new arrangement with the original. In this case the net result of the C_2 operation is the interchanging of the two hydrogen atoms in the water molecule. As hydrogen atoms are indistinguishable so the new arrangement is indistinguishable from the old. If one of the hydrogen atoms was replaced by deuterium (an isotope of hydrogen) the molecule would no longer have a C_2 axis. The water molecule is a very simple example and the rotation axis can be easily imagined. However, in general it is a good idea to make models for manipulation as this usually enables symmetry elements to be visualized more readily.

Another inorganic molecule with an easily visualized axis of rotation is ammonia, NH_3. The three-fold axis, C_3, is shown in Figure 3.1(b). Rotation by 120° in either direction about the three-fold axis results in a molecule of position and arrangement indistinguishable from the original.

Ethene has three mutually perpendicular C_2 axes, two in the molecular plane and one perpendicular to that plane as shown in Figure 3.1(c).

Benzene provides a more challenging example. The symmetry axes are shown in Figure 3.1(d). There are six C_2 axes all in the molecular plane; two are shown and the others can be inferred by analogy. Perpendicular to the C_2 axes is one six-fold (C_6) axis. Rotation by 60° about this axis produces an arrangement indistinguishable from the original. The six-fold axis in benzene has the highest order and is said to be the *principal axis*.

The two extreme orders for rotation axes are 1 and ∞. A first-order axis is trivial as rotation of any object through 360° about any axis leaves the object unchanged. The combination of any operations leaving the object unchanged is called the *identity operation*. This is mathematically necessary for group theory but need be of no further concern. An infinite order axis is found in any molecule or object that is symmetric about an axis. This means that any rotation, however slight, produces an object identical to the original. Ethyne, $HC\equiv CH$, shown in Figure 3.1(e) provides an example of an organic molecule with an infinite order axis.

3.2.2 Plane of symmetry (σ)

A molecule has a plane of symmetry if an imaginary double-sided mirror reflects both halves of the molecule so that the new arrangement is indistinguishable from the original. In other words a mirror plane divides a molecule into two symmetrical halves, each being a reflection of the other.

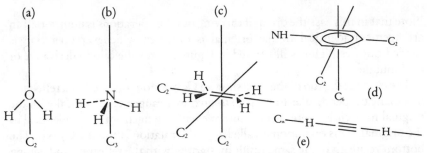

Figure 3.1 (a) The two-fold rotation axis in H_2O; (b) the three-fold rotation axis in NH_3; (c) the three two-fold rotation axes in $H_2C=CH_2$; (d) the six-fold rotation axis and two of the six, two-fold rotation axes in C_6H_6; (e) the infinite-order rotation axis in $HC\equiv CH$.

The cyclopropane derivative shown in Figure 3.2(a) is an example of a molecule that has a plane of symmetry but no axes of symmetry. For simplicity the carbon and hydrogen atoms have been omitted.

Many molecules that have planes of symmetry also have axes of symmetry. Ethene has, in addition to the three C_2 axes, three σ-planes each containing a C_2 axis and intersecting at the mid-point of the double bond. These symmetry elements are all shown in Figure 3.2(b).

Water has two mutually perpendicular σ-planes that both contain the C_2 axis and intersect along it as shown in Figure 3.2(c). Planar molecules such as water and ethene all must contain at least one symmetry plane which is the molecular plane. All linear molecules contain an infinite number of planes of symmetry containing and intersecting C_∞.

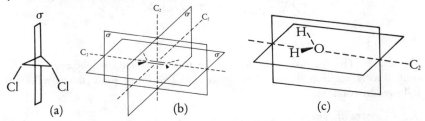

Figure 3.2 (a) The plane of symmetry in *cis*,1,2-dichlorocyclopropane; (b) the three symmetry planes in ethene; (c) the two symmetry planes in water.

When a molecule has a principal axis and one or more plane of symmetry those planes of symmetry that contain the principal axis are labelled σ_v (vertical) and those perpendicular to the principal axis are labelled σ_h (horizontal).

3.2.3 Rotation-reflection (S_n)
A combination of the two previously described operations is a distinct symmetry operation called *rotation–reflection*. This can be described as,

$$S_n = C_n \times \sigma_h = \sigma_h \times C_n$$

Note that in this case the order of carrying out the operations is immaterial; an arrangement identical to the original is obtained whichever operation is carried out first. This is illustrated in Figure 3.3 for the diiodo derivative of cyclobutane.

The route at the top of the page shows a C_2 rotation followed by a reflection in a plane perpendicular to C_2 (σ_h) to give an arrangement identical with the original molecule. Note that the intermediate arrangement is not identical to the original. This operation is called S_2 as the rotation is about a C_2 axis. The bottom route gives the same result by starting with the reflection and ending with the C_2 rotation. Again the intermediate is distinguishable from the original.

Further examples of S_n observations are given in the exercises where the use of models is recommended.

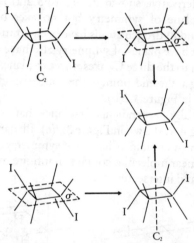

Figure 3.3 The rotation–reflection axis in *trans*-1,3-diiodocyclobutane. The top route shows rotation, followed by reflection. The lower route shows reflection followed by rotation.

3.2.4 Centre of symmetry

A molecule has a *centre of symmetry* if there is a point within the molecule such that reflection of *all* atoms through that point gives a molecule of appearance identical to the original. This operation is called *inversion* (*i*). In other words if a line drawn through the centre of symmetry meets an atom a certain distance from the centre then it will meet another identical atom on the other side of the centre at an identical distance. Mathematically this can be expressed as; if the centre of symmetry is taken as the origin then for every atom of coordinates x, y, z there is an identical atom of coordinates $-x$, $-y$, $-z$.

There are many examples of molecules that contain centres of symmetry. The derivative of ethene, shown in Figure 3.4(a) has a centre of symmetry midway along the carbon–carbon double bond. The cyclobutane derivative in Figure 3.4(b) is quite uncommon in that its *only* symmetry element is a centre of symmetry.

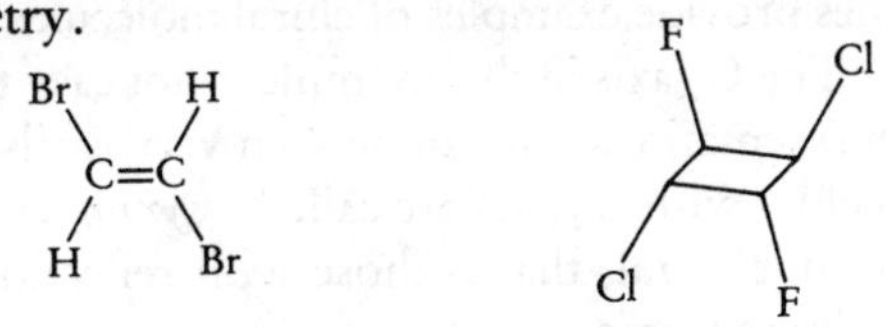

Figure 3.4   Two molecules with centres of symmetry.

3.3 Reflection symmetry

A fundamental property of molecules is their ability or inability to be superimposed on their mirror images. When a molecule is reflected in a mirror the criterion for superimposability is that the image could *exactly* fit the space occupied by the original molecule. It is clear from an inspection of Figure 3.5 that a molecule such as methane has reflection symmetry. In fact, *any* molecule with an *internal* mirror plane has reflection symmetry.

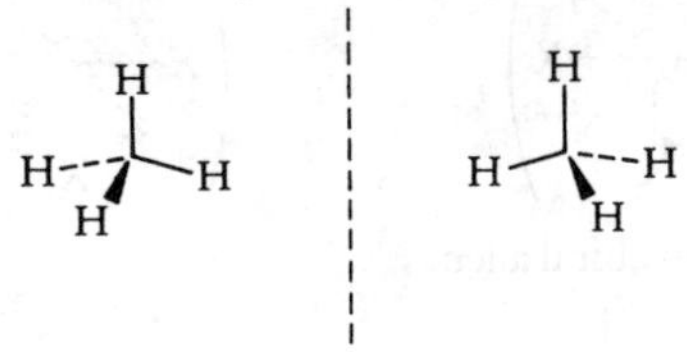

Figure 3.5   Methane has Reflection symmetry.

A molecule such as bromochlorofluoromethane shown in Figure 3.6 does not have reflection symmetry. The image and object molecules are distinct molecular species. Molecules that are not superimposable on their mirror images are called *chiral*. The word chiral is taken from the Greek word for a hand. A left hand and a right hand are, in appearance, non-superimposable. A right hand viewed in a mirror appears to be a left hand. The analogy with hands is often carried through to molecules. Chiral molecules are often said to have a 'handedness'. There are problems with the analogy but as long as only the external appearance of hands is considered the analogy is good.

The molecule shown in Figure 3.6 has no elements of symmetry whatever, and these chiral molecules are called *asymmetric*.

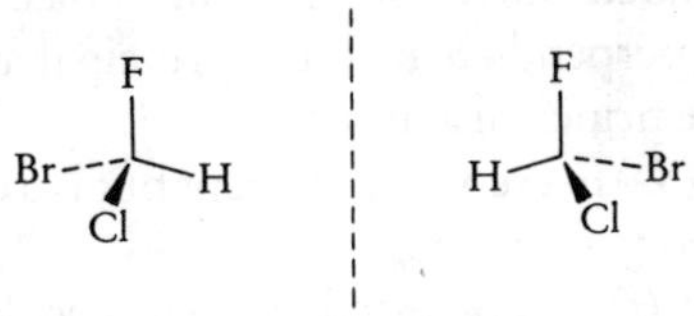

Figure 3.6   A molecule with no symmetry elements, that does not have reflection symmetry.

33

It is not a necessary condition for chirality that a molecule contains no symmetry elements. A molecule containing a C_2 axis and no mirror planes is also chiral. The allenes provide examples of chiral molecules with C_2 axes, as shown in Figure 3.7. The C_2 axis in this example is not easy to visualize from two-dimensional representations but can be seen very easily with molecular models. Chiral molecules with C_2 axes are called *disymmetric*.

Molecules that are not chiral, that is those with reflection symmetry are called *achiral* or *nondisymmetric*.

The chemical consequences of chirality will become apparent in the next Chapter, and further on in Chapter 8.

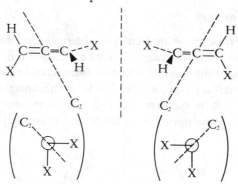

Figure 3.7 The C_2 axis in chiral allenes.

3.4 Summary of the main points of Sections 3.1–3.3

1. A molecule contains *symmetry elements* when certain parts of the molecule can be interchanged with other parts of the molecule so that the final appearance of the molecule is identical with the original.
2. The *symmetry operations* are those ways of interchanging parts of a molecule.
3. The basic symmetry elements are;
   axis of symmetry (C_n) which describes the behaviour of the molecule on rotation about an imaginary axis. The *order, n*, is given by the relationship that a rotation of $360°/n$ produces a molecule indistinguishable from the original. The *principal axis* is the axis with highest n.
   plane of symmetry (σ) describes the molecular behaviour on reflection through a double-sided mirror so that the molecule retains its original appearance. Planes perpendicular to the principal axis are labelled σ_h and those containing the principal axis σ_v.
   rotation–reflection (S_n) is a separate element but is a combination of C_n and σ_h so that $S_n = C_n \times \sigma_h = \sigma_h \times C_n$.
   centre of symmetry (i) is possessed by a molecule when the centre of symmetry is the origin and for every atom of coordinates x, y, z there is an identical atom at $-x$, $-y$, $-z$.

4. Molecules that are superimposable on their mirror images have *reflection symmetry*. A sufficient condition for reflection symmetry is that a molecule contains a mirror plane. Molecules with mirror symmetry are called *achiral* or *nondisymmetric*.
5. Molecules that are not superimposable on their mirror image are called *chiral*. Chiral molecules with no symmetry elements are called asymmetric. Those chiral molecules containing at least one symmetry element such as C_n axis are called *nondisymmetric*.

3.5 Point groups

Molecules can be grouped together in terms of the total complement of their symmetry elements. The group of all molecules with the same symmetry elements is called a *symmetry point group*.

For example, the water molecule belongs to the group of molecules (or objects) that contain one two-fold axis of symmetry and two planes of symmetry. This information is compressed into a symbol $\mathbf{C}_{2v}$. The $\mathbf{C}_2$ tells us that the axis of highest symmetry is two-fold and the v tells us that the planes of symmetry contain the axis of highest symmetry. Molecules with the same point group do not necessarily have the same appearance. The molecule CH_2Cl_2 also has the $\mathbf{C}_{2v}$ point group.

The simplest way to determine the point group of a molecule is to work logically through a set of questions concerning the number and type of symmetry elements contained in that molecule. The set of questions is reproduced in the flow chart in Figure 3.8 (p. 36).

Before any examples are attempted it is necessary to introduce three special groups for molecules of high symmetry.

Figure 3.9 (a) An inorganic octahedral complex, SiF_6^{2-}. (b) Cubane, an organic octahedral molecule.

The first is the group $\mathbf{T}_d$ that applies to regular tetrahedral molecules such as methane, carbon tetrachloride and similar molecules. A tetrahedral molecule belonging to the $\mathbf{T}_d$ group has four C_3 axes, three C_2 axes and six σ-planes. However this point group can normally be assigned by inspection of the molecule.

The other main group of special interest to chemists is the octahedral point group $\mathbf{O}_h$. An octahedral $\mathbf{O}_h$ molecule contains three C_4 axes, four C_3 axes, six C_2 axes and 9 σ-planes! In inorganic chemistry octahedral species such as $SiF_6^=$ shown in Figure 3.9(a) are common. Organic chemistry contains few examples of octahedral molecules. Cubane, shown in Figure 3.9(b) is one

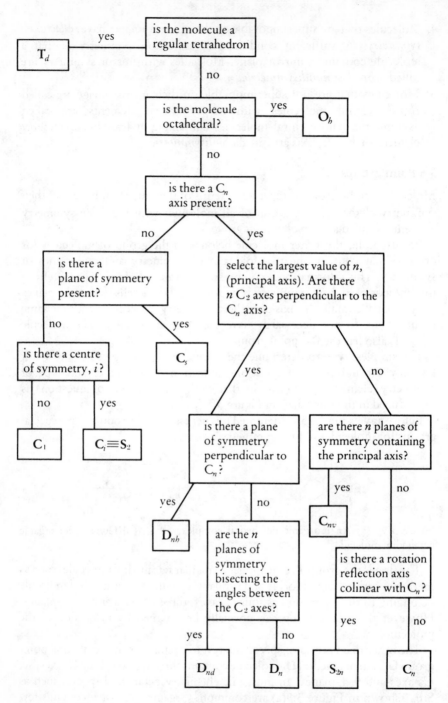

Figure 3.8   Flow chart for determining the point group of molecules.

example of an octahedral organic molecule. The hydrogen atom attached to each carbon atom is omitted for clarity. The three C_4 axes are coincident with the x, y and z axes drawn on the figure. The four C_3 axes intersect C_1 and C_7; C_2 and C_8; C_3 and C_6 and C_4 and C_5. The nine σ-planes pass through the xy, yz and xz planes; $C_2C_3C_6C_8$ and $C_1C_4C_5C_7$ and $C_1C_2C_7C_8$ and $C_3C_4C_5C_6$ and $C_1C_3C_6C_7$ and $C_2C_4C_5C_8$. The use of models may well help to visualize these symmetry elements.

The final high-symmetry point group is $\mathbf{K}_h$ which applies to objects containing *all* symmetry elements. Molecules cannot have $\mathbf{K}_h$ symmetry and in chemistry this point group applies only to single, isolated atoms. This point group can be omitted from the discussion of molecules.

All other point groups contain fewer symmetry elements than $\mathbf{T}_d$, $\mathbf{O}_h$ and $\mathbf{K}_h$. Figure 3.10 shows some simple organic molecules. The point groups of these molecules will now be determined using Figure 3.8.

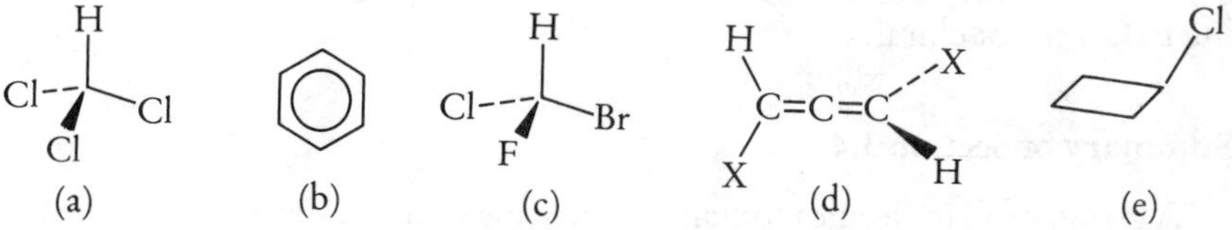

Figure 3.10 Some simple organic molecules.

The first molecule that we shall examine is chloroform, $CHCl_3$, shown in Figure 3.10(a). It is advisable to make a model of each of these molecules for inspection before working through the flow chart.

Each question will be answered in turn to obtain the point group of chloroform. The molecule is *not* a regular tetrahedron. All four groups or atoms around the central carbon atom must be indistinguishable to fulfil this condition.

The molecule is not octahedral.

There is one C_3 axis present, that passes through the C and H atoms.

Following the yes branch, the next question asks for 3 C_2 axes perpendicular to the C_3 axis. The only C_n axis in chloroform is the C_3 axis so the 'no' branch is followed.

The next question asks for n symmetry planes containing the C_n axis. Chloroform does have three planes of symmetry, each one passing through the carbon and hydrogen atoms and one chlorine atom. The 'yes' branch assigns chloroform to the $\mathbf{C}_{3v}$ point group.

The next molecule, Figure 3.10(b) is benzene. It is not tetrahedral or octahedral. There is a C_6 axis (principal axis) and six C_2 axes perpendicular to the principal axis. Following the 'yes' branch the next question asks for a plane of symmetry perpendicular to the principal axis. The molecular plane, containing all the atoms in benzene is perpendicular to C_6 and is a σ-plane. Benzene therefore belongs to the $\mathbf{D}_{6h}$ point group.

37

For the next example we return to CHClBrF (Figure 3.10(c)) which we know to be chiral from the previous section. Inspection of the figure, or a model, shows the *absence* of, a C_n axis, a plane of symmetry and a centre of symmetry. This molecule therefore belongs to the C_1 symmetry group. All molecules belonging to C_1 are chiral and, because of the absence of any symmetry elements are called *asymmetric*.

Not all chiral molecules belong to C_1 as shown by the next example, the disubstituted allene, shown in Figure 3.10(d). This molecule has one C_2 axis and no planes of symmetry. There is no S_4 axis although it may need careful manipulation of models to confirm the absence of rotation–reflection symmetry. The allene molecule therefore belongs to the C_2 point group. All molecules belonging to the C_n point groups are chiral.

The final example in this chapter is the cyclobutane derivative shown in Figure 3.10(e). There are no C_n axes in the molecule shown but there is a plane of symmetry assigning the molecule to the C_s point group. Molecules belonging to C_s are not chiral.

Summary of Section 3.4

1. A group of all molecules containing the same symmetry elements is called a *symmetry point* group.
2. The flow-chart shown in Figure 3.8 can be used to determine the point group of any molecule.

Problems and Exercises

1. Assuming that the cyclobutane ring is planar and square how many axes of symmetry can you find? Label each axis on a diagram and denote the principal axis.
2. How many axes of symmetry has iodoethyne $IC\equiv CH$ and what is the order of any axes?
3. Draw in the symmetry planes on a diagram of a dichloromethane CH_2Cl_2 molecule.
4. Which of the following molecules have S_4 axes, methane, CH_4, allene $H_2C=C=CH_2$ and ethene $H_2C=CH_2$?
5. Which of the following molecules has a centre of symmetry, i, benzene C_6H_6, ethene $H_2C=CH_2$, methane CH_4, dichloromethane CH_2Cl_2 and cyclopropane Δ?
6. Make models of HCClBrF: $ClHC=CHC$, $Cl\diagdown\square$ and

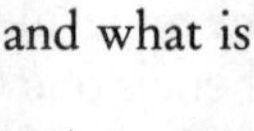

   $ClHC=C=CHCl$—exactly as shown in Figure 3.10. Then hold these models up to a mirror and make a model exactly as you see the mirror image: which molecules are chiral? What symmetry elements do the chiral molecules possess?

CHAPTER 4

Molecular isomerism—I

4.1 Introduction

The intellectual framework for understanding more complex molecular structures has now been established through; a description of the geometrical requirements of atoms in molecules, based on coordination number; a simple view of chemical bonding and, an examination of the symmetry properties of molecules.

This chapter is concerned with the classification of stereochemical relationships between pairs, or groups, of molecules *with the same molecular formula*. Most molecular formulae, for example C_9H_{20}, contain no intrinsic information on the arrangement in space of the constituent atoms. For very simple molecules there may be a unique geometry associated with the molecular formula. For example, CH_4, CH_3I, CH_2I_2, CHI_3 and CI_4 are all tetrahedral molecules with only one possible arrangement of ligands around the central carbon atom. In general, a structure can only be specified when the three-dimensional coordinates of all atoms are defined.

Isomerism concerns relationships between molecular structures, with the same molecular formulae, and is based on the symmetry properties of the molecules. Molecules with the same molecular formula but different arrangements of the atoms in space are called *isomers*. This chapter establishes the criteria for defining isomeric relationships.

It is valid to question the usefulness of any classification system. Stereochemical classifications bring an understanding of fundamental chemical and biochemical reactions, especially in understanding the, sometimes enormous, differences in reactivity between apparently closely related compounds. Additionally, one major preoccupation of organic chemists is the synthesis in the laboratory, of molecules found in small amounts in living matter. Before a complex, many step, synthesis can be embarked upon, it is essential to understand fully the stereochemistry of the compound in question. Only then can a replica of a naturally occurring compound be synthesized.

4.2 Homomers and isomers

The first condition to be satisfied in establishing isomeric relationships between structures is to confirm that the two, or more, structures have identical molecular formulae. Once this has been established the criteria of *superimposability* (or sometimes, more simply, superposability) is applied. If two molecules could, at different times, occupy exactly the same position in space they are identical and are called *homomers*. In other words, if a mould could be made of one molecule and another molecule fitted that mould exactly, then the two are superimposable and are called homomers.

If the criterion for superimposability is not met for molecules with the same molecular formula then the relationship between them is *isomeric*.

Let us examine these criteria for a few sample structures. Figure 4.1 shows two structures of formula CH_2Cl_2. It is not immediately apparent from the diagram, but examination of molecular models shows that the structures are superimposable and are therefore homomers.

The two structures in Figure 4.2 have identical molecular formulae and a superficial examination is sufficient to show that they are not superimposable as their shape is clearly different. These two compounds are isomers.

The two structures in Figure 4.3 both have the formula CHBrClF. In this case construction of models shows that however the two are orientated they are not superimposable, although the two structures are superficially very similar. So structures 4.3(a) and 4.3(b) are isomeric.

Similarly, there is an isomeric relationship between structures 4.4(a) and 4.4(b). The double bond between the two carbon atoms confers a rigidity on the molecules illustrated, as explained in the previous chapter. Molecular models of each of structures 4.4(a) and 4.4(b), if constructed properly, show that the two are not superimposable. The pairs of molecules 4.2–4.4 illustrate pairs of isomers but each pair exhibits a different type of isomerism. We shall now examine each pair more closely and define in more precise terms the various kinds of isomerism.

4.3 Constitutional isomerism

Once an isomeric relationship has been established between structures, the next step is to examine the *molecular constitution* of each structure. Molecules with the same molecular constitution have the same atoms joined together in the same order. When two organic molecules have the same molecular constitution each carbon atom in one molecule has a corresponding carbon atom in the other molecule *bearing the same atoms or groups*. However, molecules with the same molecular formula but *different* molecular constitutions are called *constitutional isomers*.

The pair of isomers in Figure 4.2 are examples of constitutional isomers. Both structures have a benzene ring with one bromine and one chlorine atom as substituents, but the molecular constitution is different. If the carbon bearing the bromine atom is numbered as C-1 the chlorine is attached to C-2 in Figure 4.2(a) but to C-4 in Figure 4.2(b); therefore each carbon atom in Figure 4.2(a) does not bear the same substituents as the corresponding carbon atoms in Figure 4.2(b).

Sometimes constitutional isomerism is further classified so that compounds 4.2(a) and 4.2(b) would be called *positional isomers*, as the positions of the substituents differ. This is not an especially useful classification except that it may be convenient in establishing the relationship between compounds that could arise from a common source. Both 2-bromochlorobenzene and 4-chlorobromobenzene would be expected as products of the bromination of chlorobenzene. Other examples of positional isomers are 1-propanol, $CH_3CH_2CH_2OH$ and 2-propanol, $CH_3CH(OH)CH_3$.

Compounds with no real chemical similarity can be constitutional isomers. The three compounds shown in Figure 4.5 are 3-methylbutanal(a), cyclopentanol(b) and propyl vinyl ether(c) and all have the formula $C_5H_{10}O$. They are constitutional isomers but belong to three dissimilar functional groups.

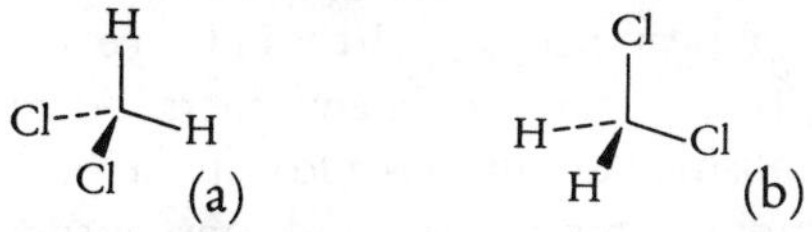

Figure 4.1   Two representations of dichloromethane.

Figure 4.2   Isomeric bromochlorobenzenes.

Figure 4.3   Isomeric bromochlorofluoromethanes.

Figure 4.4   Isomeric 1,2-dichloroethenes.

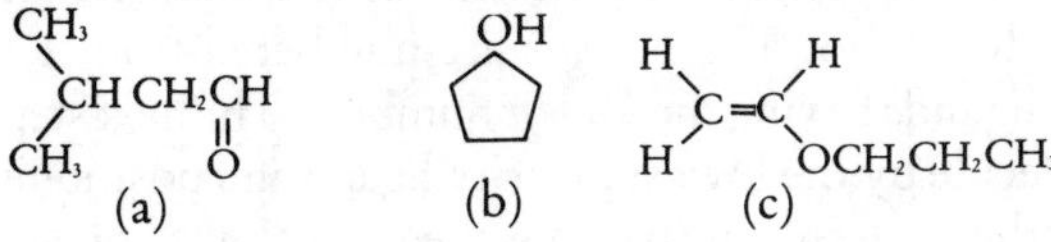

Figure 4.5   Constitutional isomers (a) 3-methylbutanal; (b) cyclopentanol; (c) propylvinylether.

The most interesting, and useful, relationships in stereochemistry are between non-superimposable structures with the *same* constitution. These are called *stereoisomers* and are molecules that differ in the arrangement in space of atoms or groups within molecules. Stereoisomerism is introduced in the next section.

4.4 Stereoisomerism

Having established that a pair of molecules have the same constitution and are non-superimposable the next step is to determine whether they are *non-superimposable mirror images*. Two chiral molecules (see Chapter 3) that are as object and mirror image cannot be superimposed and are called *enantiomers*. A simple example is the pair of structures in Figure 4.3. Both molecules can be described as, one carbon atom bearing a hydrogen atom, a chlorine atom, a bromine atom and a fluorine atom—so they have the same constitution. Making models of 4.3(a) and 4.3(b), exactly as shown in the figure, will provide a convincing demonstration that, however orientated, they cannot be superimposed. If a model of 4.3(a) is held up to a mirror the image will be identical in appearance to 4.3(b) and vice versa. The two compounds 4.3(a) and 4.3(b) are therefore enantiomers. At present we are simply concerned with establishing definitions and criteria for different types of isomerism. The more complex implications and consequences of chirality and enantiomeric relationships are covered in Chapter 8. One point that does arise at this point is how to specify the arrangement of ligands characterizing 4.3(a) and 4.3(b), and other pairs of enantiomers. This problem can be illustrated by reference to the compounds in Figure 4.3 again. Both compounds are called bromochlorofluoromethane, which does not distinguish between them, although we have shown that they are not identical. The only difference between these compounds is their *configuration*, which is the relative ordering of the atoms in space. The method chosen for specification of configuration relates the arrangements of ligands to an external chiral framework that can be understood in terms of 'left and right' or 'clockwise and anticlockwise'. In the example of the enantiomers of bromochlorofluoromethane the source of chirality is a single carbon atom, the *chiral carbon atom*, and it is necessary to specify the arrangement of atoms around that carbon atom. The configuration is specified by assigning 'priorities' based on atomic number, to the ligands around the chiral carbon atom. The details are to be found in Appendix 1 at the end of the book but the method is outlined briefly here also. The ligands are numbered from 1 to 4 with the highest priority ligands having the *lowest* numbers. The next step is to imagine holding the molecule by the lowest priority ligand and positioning this ligand at the 'back' of the molecule away from the eye. Now, you are looking at a molecule with three ligands (1, 2 and 3) pointing towards your eye, and the lowest priority ligand is directly behind the chiral carbon atom. Starting with the highest priority ligand, 1, your eye will travel either clockwise or anti-clockwise as you view the three atoms in *decreasing* priority order, 1, 2 then 3. If the eye travels clockwise the configuration is designated *R* (Rectus) and if the eye travels anticlockwise the configuration is designated *S* (Sinister). The process is illustrated in a simple, schematic way in Figure 4.6 for *R*-bromochlorofluoromethane. As with many other stereochemical problems model making will, in most cases simplify the procedure and obviates the need for difficult drawings in more complex examples.

The necessary condition for chirality is that a molecule does not have reflection symmetry (i.e. belongs to the point groups C_1, C_n or D_n). It is neither a necessary nor sufficient condition for chirality that a molecule contains chiral carbon (or other) atoms. An example of a molecule containing two chiral carbon atoms which is itself achiral is shown in Figure 4.7. That molecule has a mirror plane and therefore has reflection symmetry. This aspect of chirality is discussed in detail in Chapter 8.

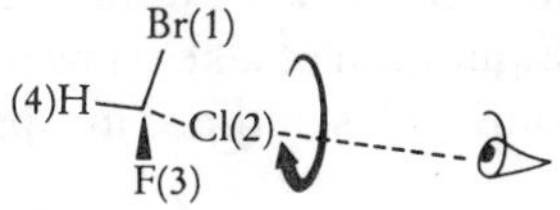

Figure 4.6   The operation of the sequence rule for *R*-bromochlorofluoromethane. The ligands are numbered from 1 to 4 in decreasing priority.

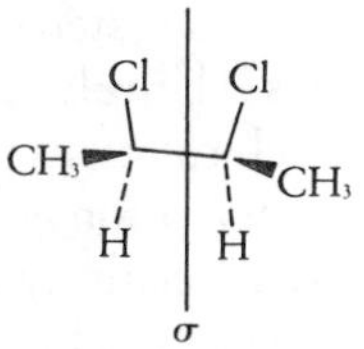

Figure 4.7   An achiral molecule that contains two chiral carbon atoms, and an internal mirror plane, σ.

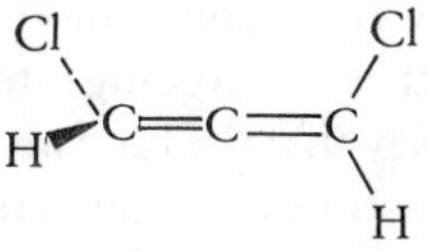

Figure 4.8   A chiral compound, dichloroallene, that contains no chiral centres.

Figure 4.9   Hexahelicene, a chiral molecule by virtue of its helicity.

A molecule can be chiral even without a chiral atom. 1,3,-Disubstituted allenes are chiral, as established in the previous chapter. A model of the dichloroallene shown in Figure 4.8, and its mirror image will demonstrate this.

Some molecules are helical in character and are chiral by virtue of their 'right' or 'left-handed' spirals. The molecule shown in Figure 4.9 is chiral and again has no chiral atoms. A model demonstrates the necessary non-planarity of this compound.

The final category in the classification of isomers concerns those molecules with the same molecular constitution that are *not* non-superimposable mirror images. Returning to the structures in Figure 4.4(a) and 4.4(b), it can readily be observed that they have the same constitution.

Each molecule may be described as having two carbon atoms, each one bound to the other and each bearing a hydrogen atom and a chlorine atom as substituents. Both molecules have at least one σ-plane and are therefore superimposable on their mirror images. Inspection of the structures shows that they are not as object and image. Molecules with this relationship are said to be *diastereoisomeric*. Compounds 4.4(a) and 4.4(b) are *diastereoisomers*.

Other examples of diastereoisomers will be met in later chapters. Alkenes may have diastereoisomeric relationships with other alkenes when each sp^2 carbon bears two different substituents. The substituents on one carbon may be the same as on the other carbon, as in 4.4(a) and 4.4(b) or may be different. The specification of configuration of alkenes uses a similar priority system to that for chiral carbon atoms and is outlined in Appendix 1.

4.5 Molecular motion

So far we have concentrated on stereoisomeric relationships between molecules in which the shape is relatively time-independent. A question that arises concerns the way in which rotations about bonds affect stereochemical relationships. Another is how stereoisomeric relationships can be determined for compounds in which there is relatively free rotation about single bonds. The molecule shown in Figure 4.7 has been given a particular structure but rotation is expected about the central (2,3) carbon–carbon bond. The next chapter answers these questions and demonstrates the stereochemical methodology applied to molecules undergoing internal molecular motions. For the moment it is sufficient to say that the fundamental definitions of enantiomers, diastereoisomers and constitutional isomers are still valid although extra provisions must be made to complete them.

4.6 Conclusion and summary

The various stereochemical relationships have been outlined in Sections 4.1–4.4. Each section builds on the last and a logical sequence of symmetry arguments is used to determine the relationships between structures. The definitions are all contained within the text. Instead of the usual summary we have summarized this chapter as a flow-chart which should enable you to distinguish between the different types of isomerism. Before you look at the chart (Figure 4.10) it would be a useful exercise to read through the text again and attempt to make one for yourself. Try to frame a series of questions with yes or no answers in such a way that the relationships between two compounds may be traced through from no stereochemical relationship to a diastereoisomeric relationship.

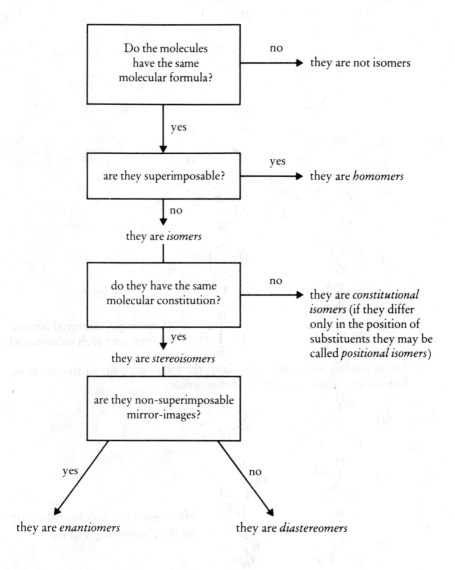

Figure 4.10   Flow chart for determining the stereochemical relationships between molecules.

Problems and Exercises

1. Using models where necessary label the following pairs of compounds as homomers, constitutional isomers, diastereoisomers or enantiomers.

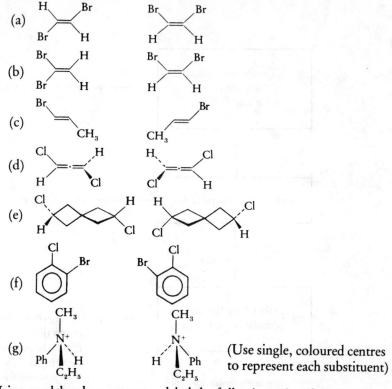

(a)

(b)

(c)

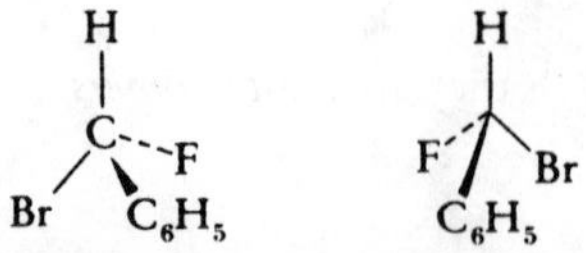

(d)

(e)

(f)

(g)   (Use single, coloured centres to represent each substituent)

2. Using models where necessary label the following pairs of structures as, homomers, diastereoisomers or enantiomers.

(a)

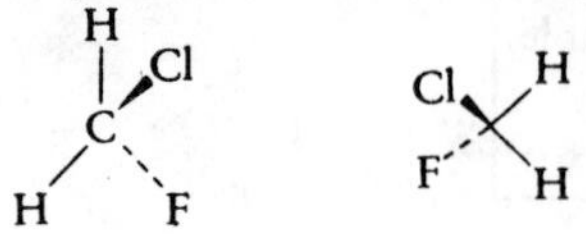

(b)    Note: use a single coloured centre for the phenyl ring, C_6H_5.

(c)

What does this exercise tell you about the number of groups or atoms that must be arranged around a single carbon atom before that carbon atom can be chiral?

3. Label the following pairs of structures as enantiomers, diastereoisomers or homomers. Which molecules are chiral?

(a)

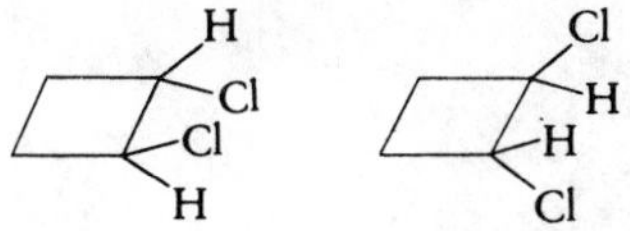

(b)

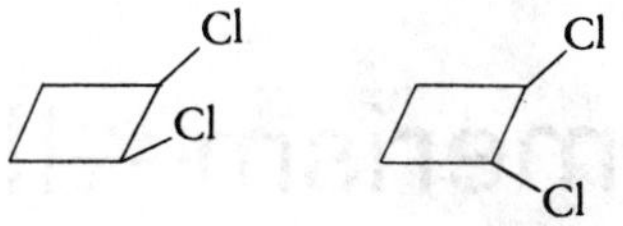

(c)

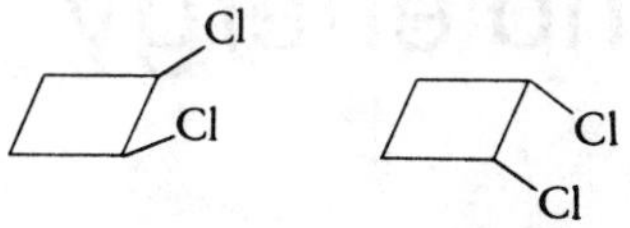

(d)

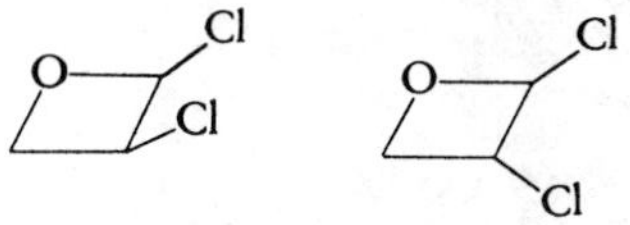

CHAPTER 5

Molecular isomerism—II. Time scales and energy criteria

5.1 Introduction

The time has come to confront the problems of the stereochemical consequences of motion within molecules. The concept of isomerism is extended in this chapter to account for the effects of intramolecular motion.

So far, it has been relatively easy to compare two molecules. They are identical if they can be superimposed, and a pure chemical compound would consist of an ensemble of molecules of identical composition and geometry. But in Chapter 2 it was shown that there can be rotations about single σ-bonds. This immediately changes our view of chemical compounds as it implies that molecular geometry is not fixed; it is *dynamic*. If we take a very simple example, ethane, some of the problems can be illustrated. Make a simple model of ethane and hold one methyl group in one hand and rotate the other about the C—C bond rather like a propeller. By doing to you are changing the geometry of the ethane molecule. There are an infinite number of geometries possible for ethane (albeit very slightly different), so how can we compare the geometries and what effect does this dynamic behaviour have on structure? To answer these questions we need to know more about the process of rotation and the energies of different geometries. Using ethane, and its derivatives as examples refined definitions and criteria can be proposed.

As we shall be concerned with energy criteria both here and in subsequent chapters it is worthwhile spending a short time defining some of the energy terms used. The Gibbs free energy (sometimes Gibbs energy), difference ($\triangle G^{\ominus}$) between two species in equilibrium is related to the equilibrium constant, K, by,

$$\triangle G^{\ominus} = -RT \ln K$$

where R is the gas constant and T is the temperature of measurement. The Gibbs free energy of a reaction is a composite energy term and should not be confused with the enthalpy ($\Delta H^{\ominus}$) of a species. The enthalpy difference term is a measure of the energy in the form of heat, either liberated or absorbed in the interconversion of two molecular species. Enthalpy and Gibbs free energy are related by

$$\Delta G^{\ominus} = \Delta H^{\ominus} - T \Delta S^{\ominus}$$

where $\Delta S^{\ominus}$, the *entropy* difference, is a statistical term related to the distribution of energy throughout the system, and the symmetry properties of the species. Unless otherwise stated it may be assumed that the term energy, used unqualified, refers to Gibbs free energy. A fuller description is given in Chapter 10.

5.2 Conformation

The solution to the stereochemical classification dilemma starts with the knowledge that there is only one *compound* ethane, but apparently an infinite number of ethane geometries. (Although in all geometries the same atoms are bonded together by the same kinds of bonds.) It is not an easy matter to define a chemical compound, although it is a fundamental chemical concept. One definition is that a chemical compound is something that can, in principle, be isolated as a pure substance with a defined set of physical properties. All the molecules in a compound must have the same chemical constitution.

Before the dilemma can be completely resolved we must look more closely at bond rotations. The molecules that we have examined so far have been defined by their constitution, bond lengths, bond angles and configuration. It is now important to add a further parameter, the *dihedral angle* which defines the degree of rotation about a particular bond in a particular geometry of a molecule. An elegant way of showing molecular geometry, including rotations, is the *Newman projection* which is shown for ethane in Figure 5.1.

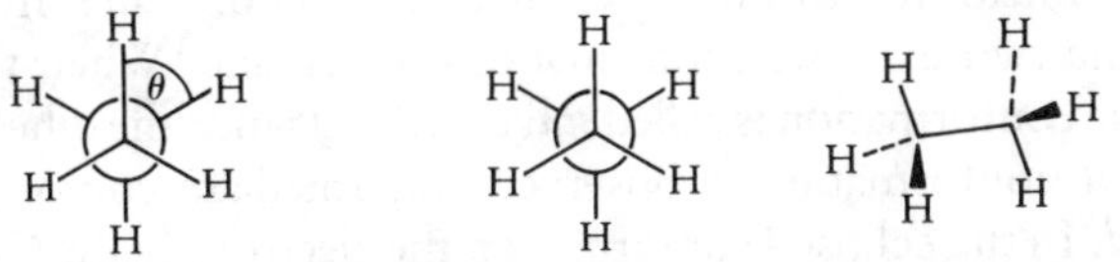

Figure 5.1 A Newman projection of ethane, in which the dihedral angle, θ, is defined.

Newman projections are representations of geometry looking along a particular bond. The atom nearer the eye is shown as a point and the three ligands attached to that forward atom are joined by solid lines meeting at that point. The configuration about that forward atom is fixed and, if all the ligands are different, defined by their clockwise or anticlockwise disposition according

49

to the sequence rules. The rear atom is immediately behind the near atom and the configuration of the ligands attached to that farther atom is shown by their attachment to a circle centred on the internuclear axis. The angle marked on Figure 5.1. as θ is called the dihedral angle. Varying the dihedral angle changes the molecular geometry, so that the effect of rotation about bonds can be illustrated.

Now we can completely specify the geometry of a molecule by including the dihedral angles. A *conformation* is defined as a particular geometry of a molecule that is specified by indicating all bond angles, bond lengths, configurations (where appropriate), and dihedral angles. A molecule has, at any given time, one conformation although its conformation may change with time.

5.3 Conformational changes and energy barriers

Ethane molecules are constantly changing conformation. Any change in motion requires an energy transfer. The energy necessary for changes in ethane molecular motion is obtained from collisions between ethane molecules and other molecules or the walls of the vessel.

If all ethane conformations were of equal energy there would be an equal probability of finding any of the possible conformations. This is not the case, all molecules exhibit *conformational preferences*. This is covered in detail in the next chapter but must be introduced briefly here to enable the new criteria for isomerism to be developed.

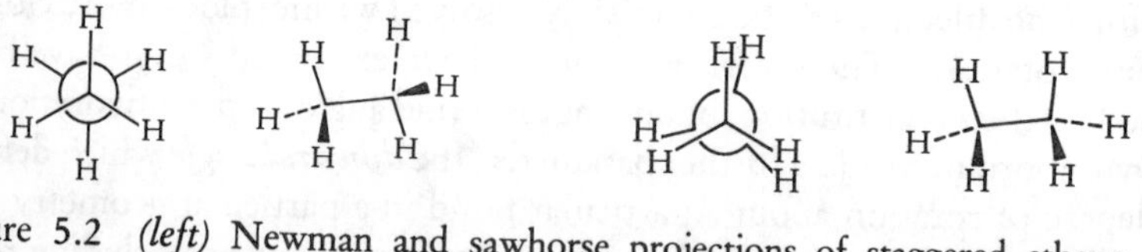

Figure 5.2 *(left)* Newman and sawhorse projections of staggered ethane, with $\theta = 60°$. *(right)* Newman and sawhorse projections of eclipsed ethane, with $\theta = 0°$.

Figure 5.2 illustrates through Newman projections and flying wedge projections the two extreme conformations for ethane. When the dihedral angle is 60° the conformation is called *staggered* and when the dihedral angle is 0° the *eclipsed* conformation is observed. Intermediate conformations are called *skewed*. In the eclipsed conformation the electrons in the C—H bonds on different carbon atoms are as close together as possible, given the constraint of fixed bond angles and lengths. On the other hand, in the staggered conformation, the C—H bond electrons are as far away from each other as possible. It is now well-established that the *staggered* conformation is of lower energy than the eclipsed conformation. The change in energy of an ethane molecule as the dihedral angle is changed is illustrated in Figure 5.3. This shows the staggered conformations at energy troughs and the eclipsed conformations at peaks. Energy needs to be added to a molecule to convert

50

one staggered conformation into another. The energy barrier for ethane is quite low, 12 kJ mol^{-1}, which means that at room temperature the ethane conformation is changing extremely rapidly. The motion is not quite like that of a smoothly running propellor. The ethane molecule resides in a staggered conformation for a finite, but very short *residence time*, τ seconds, and then very rapidly compared with τ, changes to another staggered conformation, and so on. Measurement techniques with relatively long time scales would show an averaged geometry for ethane, but very short-timescale techniques could observe ethane in the staggered conformation. As with any process with an energy barrier, ethane rotation is faster at higher temperatures and slower at lower temperatures.

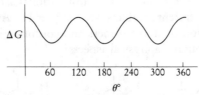

Figure 5.3 The variation of the Gibbs free energy of ethane with changes in the dihedral angle.

In general, physical measurement techniques only allow observation of conformations at energy minima. The residence time in conformations not at energy minima is too short for measurement. For open-chain compounds, staggered conformations are usually at energy minima and eclipsed conformations at energy maxima.

Molecular conformations at energy minima are called *conformers* (a contraction of conformational isomers).

Returning to the problem of the variable geometry of ethane, and other molecules, we say that *any molecules, of the same molecular constitution, that are rapidly interconverting, at the temperature of measurement are molecules of the same compound.* This is expanded and explained in later sections.

5.4 Conformational isomerism

If it was possible to label the six hydrogen atoms in ethane there would be three distinct conformers with staggered conformations. These would then be labelled as *conformational isomers,* but as there is no way of distinguishing hydrogen atoms all three conformers are *conformational homomers.* Replacing one hydrogen atom on each carbon with deuterium does allow the distinction of conformers. Figure 5.4 (p. 52) shows the three 1,2-dideutero-ethane conformers that are separated by energy barriers. (Generally, staggered conformations are at energy minima and eclipsed conformations are at energy maxima.) Consider first Figure 5.4(a) and 5.4(b). Each of these conformers has the same molecular constitution. The only symmetry element in each is a two-fold rotation axis so both conformers are disymmetric with point group C_2. They are also non-superimposable mirror images. Applying the same criteria for comparison as used in the previous chapter it can be seen that there is an enantiomeric relationship between 5.4(a) and 5.4(b). However, these two conformers may be *readily* interchanged, by bond rotations, and are therefore labelled as *conformational enantiomers*. Similarly 5.4(a) and 5.4(b) are *conformational diastereoisomers* of 5.4(c). The definition will need to be clarified in the light of further examples but in general, conformational isomers are those which are readily interchangeable and therefore in equilibrium at the temperature of measurement. There are sound *practical* reasons for giving stereochemical labels to conformers. In chemical reactions it is quite common for different conformational isomers to give rise to stereochemically different products (see Chapter 11). Attention is now turned to the differentiation of stereoisomers and conformational stereoisomers; a problem not without metaphysical aspects!

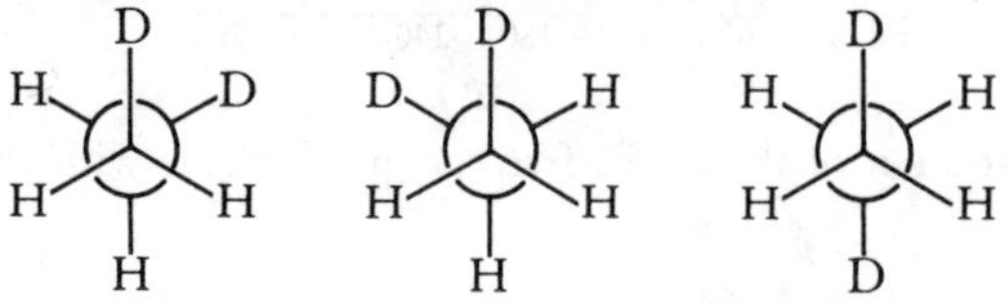

Figure 5.4   Conformational isomers of 1,2-dideuteroethane.

5.5 Conformational isomerism and isomerism

As with ethane, there is only one *compound* 1,2-dideuteroethane, but there are three distinct and, in principle, observable conformers of that compound. For two samples of 1,2-dideuteroethane we should like to say, on the one hand, that the molecules in one sample are homomers of those in the second sample, on the other hand, within either sample we would like to recognize that there are homomeric, enantiomeric and diastereoisomeric relationships between conformers.

Stereoisomerism as described in the previous chapter implies a fixed, time-invariant, relationship between two samples. For example, the structures 5.5(a) and 5.5(b) are enantiomeric and cannot be interchanged by rotations about the C—C (or any other) bonds. The *compounds* represented by structures 5.5(a) and 5.5(b) are enantiomers. A definition sometimes used, but not favoured by the author (as outlined later in the chapter) is that stereoisomers that cannot be interconverted without breaking and remaking bonds are *configurational isomers*.

But, what of the relationship between 5.5(a) and 5.5(c); and between 5.5(b) and 5.5(c)? The conformers 5.5(b) and 5.5(c) *can* be interconverted by rotation about the C—C bond and they are conformational diastereoisomers. Now let us take a hypothetical situation to illustrate the methodology.

52

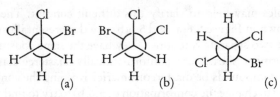

Figure 5.5 (a, b) Enantiomeric conformations of the enantiomeric compounds BrClHCCH$_2$Cl. (c) A conformational diastereoisomer of (b).

Suppose we could obtain a bottle of molecules with conformation 5.5(c) and another bottle of molecules with conformation 5.5(b). Within a fraction of a second at normal temperatures both bottles would contain, on average, the same distribution of molecular geometries, comprising a weighted-average proportion of molecules of each of the available conformations. On the laboratory time-scale 5.5(b) and 5.5(c) would be indistinguishable and therefore represent the *same compound*, because they would be inter-converting too rapidly to *isolate*. It is possible that at low temperatures, that 5.5(b) and 5.5(c) could be observed (in equilibrium with other conformations) using a fast technique such as infra red spectroscopy.

We are slowly moving towards a workable distinction between isomerism and the sub-class of conformational isomerism, based on the ability to isolate (in principle) isomers. The *macroscopic* properties of an ensemble of molecules are determined by the overall composition of conformers that are in equilibrium.

Frequently, a chemical compound is represented by just one of the con-formations available to that compound. If it is necessary to know the stereo-chemical relationship between two *compounds* each represented by one or more conformations the following procedure can be applied. First, it is necessary to know that the structures are conformationally mobile—that there is fairly rapid rotation about the bonds in question. Second, the conformation of one structure is compared with the conformation of the other. *The criterion for determining the relationship between two compounds represented by single conformations depends on the closest relationship that can be obtained between the two structures.* If two conformations are homomers or can be made so by σ-bond rotations, then the two *compounds* represented by those conformations are homomers. Similarly, if two conformations can be made enantiomeric *but not homomeric* then the two compounds represented by those conformations are enantiomeric. Two conformations that cannot be made homomeric or enantiomeric represent diastereoisomeric compounds, assuming the same molecular constitution.

When considering the relationship between pairs of *compounds* we have shown that it is not sufficient to consider only one conformation of each compound. In the case of molecules that do not change conformation, such as CHBrClF and ClHC=CHCl then the result of comparing two molecules gives the same results as comparing the overall ensemble of molecules.

53

Some examples may help to clarify this difficult concept. The Newman projections shown in Figure 5.6(a) and 5.6(b) could represent illustrations of the products of two separate reactions. They have the same constitution and the two conformers shown are conformationally diastereoisomeric, but would the two compounds be diastereoisomeric? Keeping the conformation 5.6(a) fixed we can change the conformation of 5.6(b) to try to find the closest stereochemical relationship. The first step shown in 5.6(c) is to rotate the whole of the Newman projection 5.6(b) through 120° in an anticlockwise direction to give 5.6(c). The orientation and configuration of the front carbon atoms in 5.6(a) and 5.6(c) are now enantiomeric. Rotation of the rear carbon atom by 120° in a clockwise direction shows that 5.6(a) and 5.6(b) are *enantiomers*; further rotation is unable to give rise to homomeric conformations.

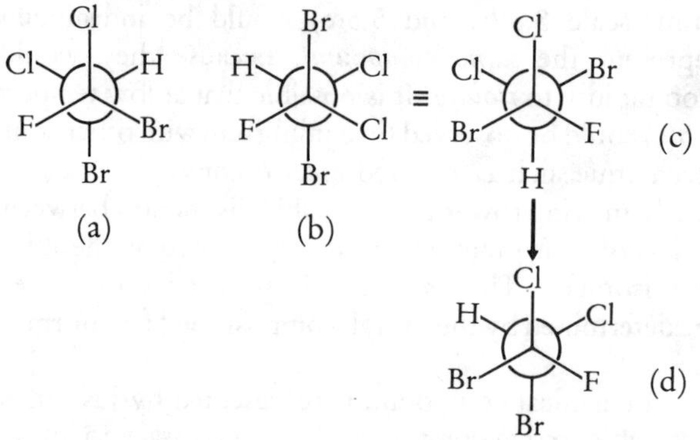

Figure 5.6 (a) and (b) are diastereoisomeric structures, but represent enantiomeric *compounds*.

The next example refers to the compounds represented by the structures shown in Figure 5.7. These two structures are enantiomeric and this is the closest relationship possible for these structures in the *staggered conformations as they are drawn*. The direction along which we look when drawing Newman projections is entirely arbitrary. It is conventional when comparing two projections to ensure that the same bond is viewed in the same direction. The structures in Figures 5.7(a) and 5.7(b) have the same groups attached to each carbon atom. In these cases it is essential to view, and compare, the Newman projections from each direction. Figure 5.7(c) shows the same structure as 5.7(b) but looked at from the other direction, this structure is conformationally diastereoisomeric with 5.7(a). Now, by rotating the front group through 60° clockwise and the rear group by 60° anticlockwise conformation 5.7(d) is obtained, which is identical with conformation 5.7(a). Therefore the compounds shown as 5.7(a) and 5.7(b) are homomers and the structures shown are conformational enantiomers.

One further, but related, problem concerns the criteria for deciding whether a *compound*, represented by one or more conformations is chiral.

54

The manifestations of chirality of a compound are a result of contributions from all available conformations of the molecules, assuming that these are readily interconvertible at the temperature of measurement. The rule, for readily interconvertible conformations, of a molecule is that a *compound cannot be chiral* if any obtainable conformation is non-disymmetric *or* if each chiral conformation is in equilibrium with its enantiomer. In most examples of achiral compounds, with chiral conformations, both of the above conditions can be met.

The compound represented in Figure 5.7 illustrates the point well. There are no non-disymmetric staggered conformations for that compound but we have already shown that enantiomeric conformations are in equilibrium (5.7(a) and 5.7(b)). The compound shown in Figure 5.7 should therefore be *achiral* by the second of the two criteria. The first criterion states that any non-disymmetric conformation of a molecule renders the compound achiral. Manipulation of any of the conformations enables the production of the eclipsed conformation shown in Figure 5.8. This conformation has reflection symmetry resulting from a mirror plane perpendicular to the C—C bond axis (point group C_s). It is irrelevant that the achiral conformation is not at an energy minimum; *it is only necessary that the conformation be passed through during the ready interconversion of other conformations.*

There are still a number of problems in definition to be overcome before stereochemical classifications can be understood completely. Some of these will be discussed in Section 5.7 but it is useful to summarize the classifications described so far.

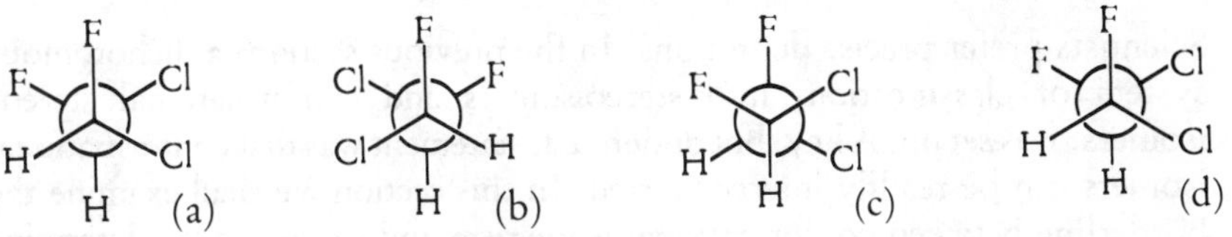

Figure 5.7   Conformations of R,S-1,2-dichloro-1,2-difluoroethane.

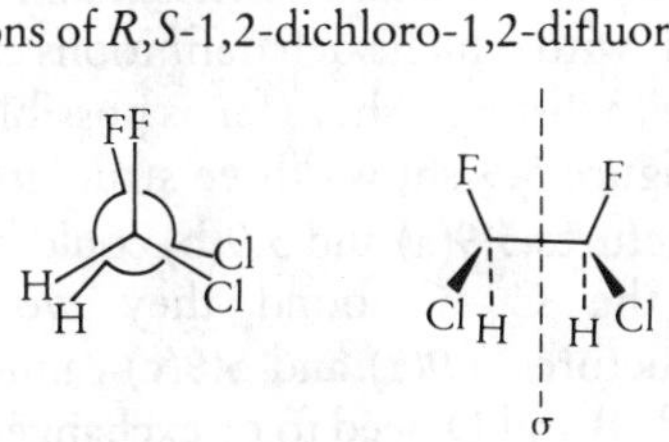

Figure 5.8   An achiral conformation of R,S-ClFCHCHFCl.

5.6 Summary of Sections 5.1–5.5

1. The conformation of a molecule is a particular geometry of that molecule that is described in terms of bond angles, bond lengths, configurations and dihedral angles. A molecule may have more than one conformation.

2. The dihedral angle θ is the angle subtended by two chosen groups when viewed along a bond about which rotation is possible.

3. When a molecule can exist in a number of readily interconvertible conformations the properties of a chemical compound made up of an ensemble of such molecules are determined by the sum and proportions of all available conformations.

4. In general only conformations found at energy minima can be observed. Such conformations, separated by energy barriers, are called conformers.

5. Conformational Isomerism describes the relationship between conformers.

6. When considering stereochemical relationships between two *compounds,* that may each be represented by more than one conformation, the relationship is determined by finding the closest possible correspondence between the conformations representing each compound. Rotation about single bonds may be necessary to find the closest correspondence. This classification is only applicable when the conformations in question are readily interconvertible at the temperature of measurement. The precedence for 'closeness of correspondence' is homomer > enantiomer > diastereoisomer.

5.7 The grey area, and what is meant by readily interconvertible

Scientists prefer precise definitions. In the previous sections a dichotomous system of classification, into stereoisomers and conformational stereoisomers, was set up. A key, but undefined, statement was that conformational isomers can be readily interconverted. In this section we shall examine the borderline between conformational isomerism and isomerism to determine the degree of precision with which such definitions can be made.

First, it is necessary to distinguish, as far as possible between conformation and configuration. Figure 5.9 shows three structures of molecular formula XYZCCABD. If structures 5.9(a) and 5.9(b) could be interconverted readily by rotation about the C—C bond they would be conformational diastereoisomers. Structures 5.9(a) and 5.9(c) cannot be interconverted by bond rotations; ligands B and D need to be exchanged in one structure before interconversion can be achieved. So far these structures have simply been labelled diastereoisomeric. More recent chemical literature has tended to label stereoisomers that have a 'high' energy barrier to interconversion as *configurational stereoisomers.*

There are two problems with these definitions. One is that we must have a clearer idea about what constitutes 'high' and 'low' energy barriers. The other is that, as described in Section 5.5, older terms imply that configurations can *only* be interchanged by breaking and making bonds and conformational changes refer to rotations about single bonds.

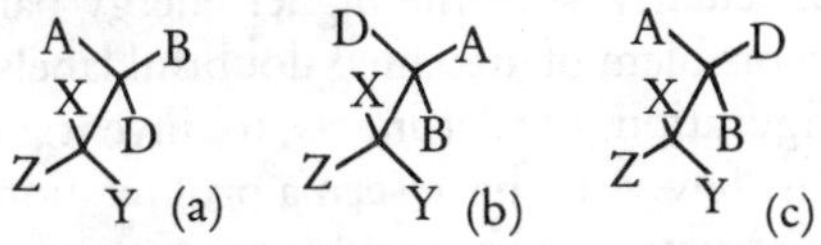

Figure 5.9   Three structures XYZCCABD.

At the moment there is no real solution to the energy barrier problem. Extreme examples can be identified as configurational or conformational isomers. For example, stereoisomers that can be isolated and characterized are called configurational isomers. Stereoisomers that can be observed spectroscopically but never separated are conformational isomers.

It can be estimated that stereoisomers with an independent lifetime of an hour or two need to have a Gibbs free energy of activation to interconversion of about 100 kJ mol^{-1}. (Corresponding to a first order rate constant of less than 10^{-4} sec^{-1}.) Therefore any stereoisomers separated by an energy barrier of more than 100 kJ mol^{-1} can be called configurational isomers. Similarly it is unlikely that stereoisomers with a barrier to interconversion of less than about 60 kJ mol^{-1} could be separated at ambient temperature. Unless an international body sets an arbitrary limit of say 80 kJ mol^{-1} as separating configurational isomers from conformational isomers, stereoisomers separated by barriers of 60 to 100 kJ mol^{-1} will remain in the 'grey area' as indefinable. Similarly compounds may be configurational isomers at 298 K but conformational isomers at 398 K.

Some examples of such problems are now presented briefly. Each type of isomerism is covered in more detail in the later chapters.

5.7.1 Amide restricted rotation

Rotations about simple single bonds are generally classed as conformational changes. The barrier to rotation about double bonds is usually significantly higher than 130 kJ mol^{-1}, and rotation involves breaking and then remaking the π-bond. Rotations about double bonds are therefore usually classed as configurational changes. Amides have, as described in Chapter 2, a partial double bond between the carbonyl carbon atom and the nitrogen atom. The amide molecular framework $-C{<}^{O}_{N}$ is planar and the three atoms bonded to the carbon and nitrogen atoms also lie in that plane. Rotations of the type shown in Figure 5.10 (p.58) can be observed readily using nuclear magnetic resonance or vibrational spectroscopy although the species 5.10(a) and 5.10(b) cannot generally be isolated in pure form. The energy barrier to rotation about the C—N bond falls in the range 60–90 kJ mol^{-1}. This barrier places amide rotational isomerism right in the 'grey area' between conformational and configurational changes. Most workers in the field of amide rotation prefer to describe the process as a conformational change although it would not be incorrect to suggest that the change was configurational, particularly with the higher energy barrier rotations. One way of avoiding the problem of attaching doubtful labels to processes on the conformation/configuration borderline is to invent new terms for such processes and the last few years have seen a proliferation of such terms. For example, any set of structures obtained by rotation about a single bond, or partial double bond are called *rotamers* (short for rotational isomers). The amide structures 5.10(a) and 5.10(b) can therefore be called rotamers, which does not imply a particular energy classification. Amide isomerism is discussed in more detail in Chapter 6.

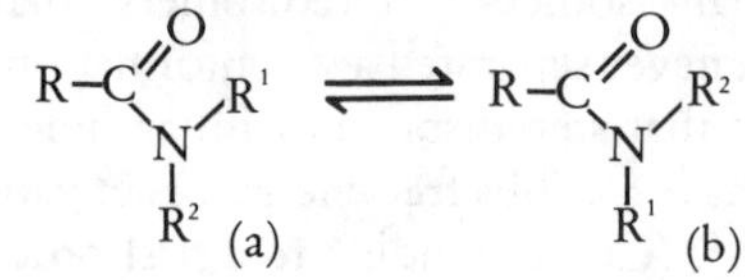

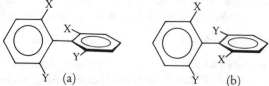

Figure 5.10   Amide rotamers.

5.7.2 Isomerism in biphenyl compounds

The chemical bond between the two benzene rings in the biphenyl compounds in Figures 5.11(a) and 5.11(b) is a simple σ-bond. However there are numerous examples where the enantiomeric structures 5.11(a) and 5.11(b) can be *isolated* as pure substances. The Gibbs energy barrier to rotation about the single bond can be as high as 140 kJ mol^{-1}. The problem here is again one of definition. No bonds are broken in the interconversion of the enantiomeric structures so by simple definitions this is a conformational change. As the enantiomeric species can be isolated as pure compounds, they are called, on an energy criterion, configurational isomers. To avoid the problem a new term has again been introduced. *Atropisomers* are species that are isolable but can be interchanged by rotation about single bonds. This definition is general and not restricted to biphenyl compounds. *Atropisomerism* refers to restricted rotation about single bonds. This topic is also examined in more depth in Chapter 6.

Figure 5.11   Biphenyl atropisomers.

5.7.3 Isomerism in trans-cyclooctene

Another example of the conformation/configuration dichotomy is given by *trans*-cyclooctene. As shown in Figure 5.12 this compound can exist in two enantiomeric forms. Each form can be isolated and the barrier to inter-conversion is about 120 kJ mol^{-1}. The two forms, 5.12(a) and 5.12(b) can be interconverted by a non bond-breaking process involving twisting the double bond through the inside of the ring. This should be verified by the use of models. It is better to attempt this first on models of the carbon framework only. The reason for the high barrier to interconversion is steric hindrance of hydrogen atoms in the molecule. If the hydrogen atoms are represented by straws it can be observed that the process is more difficult.

There does not seem to be a specific term for this type of isomerism and the safest description refers to the 'two enantiomers of *trans*-cyclooctene'.

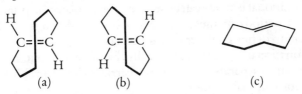

Figure 5.12   (a) and (b) are enantiomeric *trans*-cyclooctene structures. (c) Alternative representation of *trans*-cyclooctene.

5.7.4 Amine inversions

The processes described above have all concerned formally conformational changes with high energy barriers. The next two processes are formally configurational changes with low energy barriers. Amines, such as those represented by structures 5.13(a) and 5.13(b) are chiral; but, however synthesized such compounds usually exhibit the properties of an equimolar mixture of the two enantiomers. There is usually rapid interconversion of amine enantiomers with energy barriers in the range 15–130 kJ mol^{-1}. This is defined as a configurational change as the arrangement of ligands is different in each enantiomer. In fact no bonds are broken as in the change as the process proceeds through a planar transition state but the configuration nomenclature is retained as the change can be described as $S \rightarrow R$ or $R \rightarrow S$. The amine interconversion is described as an *inversion* (as in the turning inside-out of an umbrella) and to avoid misleading labels the enantiomers are called *invertomers*.

$$X\text{--}\overset{N}{\underset{Y\ \ (a)}{\diagdown}}Z \quad \rightleftharpoons \quad Z\text{--}\overset{N}{\underset{Y\ \ (b)}{\diagdown}}\text{--}X$$

Figure 5.13   Amine invertomers.

5.7.5 Facile double bond rotations

Most carbon–carbon bonds are resistant to rotation at normal temperatures. There are certain double bonds, such as those illustrated in Figure 5.14 (p. 60), where rotation is relatively rapid at normal temperatures. In these examples the configurational nomenclature is generally retained even for those rotations with energy barriers of about 60 to 70 kJ mol^{-1}.

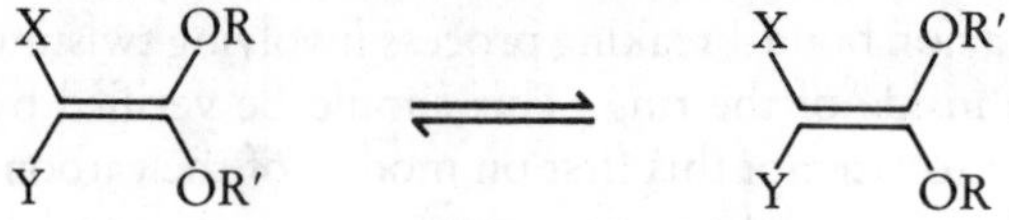

Figure 5.14   Facile double bond rotations.

5.8 Summary and conclusion of Section 5.7

1. Configurational isomers can be separated and isolated, in principle, at least for short periods of time. An energy barrier of about 100 kJ mol^{-1} is sufficient to define configurational isomers.
2. Conformational isomers can usually be observed but not separated. Stereoisomers separated by an energy barrier of less than about 60 kJ mol^{-1} are conformational isomers.
3. Where isomers are separted by barriers of about 60 to 100 kJ mol^{-1} definitions are based on structural features or new terms are introduced.
4. Rotamers are stereoisomers obtained by rotation about a single bond.
5. Atropisomers are isolable stereoisomers interchangeable by rotation about a single bond. Atropisomerism refers to restricted rotation about single bonds.
6. Invertomers are enantiomers that are readily interchanged by non-bond-breaking inversion at a particular atom (nitrogen in particular).

5.9 Physical techniques and time-scales

There are now a large number of structure-determining techniques available to the organic chemist. There is unfortunately insufficient space available to describe these in detail and you are referred to specialist texts at the end of the book. As stereochemistry is an integral part of organic chemistry it is necessary to know a little of the methods of structure determination and their limitations. This brief survey has been left until this point because time scales are a most important aspect of physical techniques. When confronted by structural data on a compound it is essential to know-how those data were obtained so that the stereochemical implications can be assessed. This is especially true for conformationally mobile molecules. It has already been pointed out that photographs of moving objects will show different aspects of the motion depending on the shutter speed. Each physical technique has a *time-constant* range which is in some ways analogous to the shutter speed of a camera. If a technique has a time constant of, say, 10^{-2} sec, then that technique will not be able to observe changes taking place in less than 10^{-2} sec. In a hypothetical example, consider a molecule with two conformers that are interchanging. If the *mean lifetime* of each conformer is less than 10^{-2} sec (i.e.

the molecule exists in a conformation for $< 10^{-2}$ sec before changing to the other conformation) then the technique in question gives a spectrum that is an average of the two conformations. On the other hand, if each conformer exists for longer than 10^{-2} sec each conformer can be observed separately.

The time constant of a physical technique is related to the frequency of the radiation used in the study. Very high frequency radiation enables fast processes to be studied as the time constant is very short. The fastest techniques are those in the X-ray region of the electromagnetic spectrum with frequencies of around 10^{12} MHz.

Table 5.1 gives a summary of the information obtained from, and the time scales of some common physical techniques. The table is divided into two sections; diffraction techniques and spectroscopic techniques. Diffraction techniques depend on the scattering of radiation (or electrons or neutrons) by atoms. Diffraction patterns are obtained from regular arrays of atoms and complex computing is usually needed to obtain structural information from these. (Table 5.1 is shown on pp. 62–63.)

Spectroscopic techniques depend on the fact that each molecule has a set of *quantized energy levels* that are characteristic of that molecule. In general molecules only absorb radiation of energy equal to the difference in energy between two energy levels (*the resonance condition*). Structural information is then deduced from the energy of the absorbed radiation. Each technique has its own strengths and weaknesses; summarized in the table. The techniques in the table have time constants varying between about 1 second and about 10^{-15} sec. Processes taking place more slowly than one second are usually studied by *kinetic measurements*. One or other of the spectroscopic techniques is used to measure the 'instantaneous concentration' of species in a sample and spectra are taken at appropriate time intervals and the changes monitored. The kinetic 'time constants' can vary from about 1 second to days, months or, with patience, years!

Table 5.1. Physical techniques of major use in stereochemistry

Technique	Wavelength of radiation/cm	Approximate time constant/sec	Information obtained and limitations
Diffraction techniques			
X-ray diffraction	10^{-7}–10^{-8}	10^{-15}	Complete structure, bond angles and lengths. Hydrogen atoms difficult to observe, usually neglected. Needs crystalline compound and expensive equipment.
Neutron diffraction		$< 10^{-15}$	Complete structure, hydrogen atoms easily observed. Crystalline sample, expensive equipment.
Electron diffraction		$< 10^{-15}$	Complete structure for simple molecules, usually gaseous sample required.
Spectroscopic techniques			
electronic spectroscopy (ultra violet and visible)	10^{-4}–10^{-5}	10^{-13}	Best used for comparison of two or more samples. Can be used to infer conjugation or electron delocalization. Solutions of samples generally used. Routine technique, extensively used for kinetic studies over long periods.

Technique		Description
Vibrational spectroscopy (infrared and Raman)	$\sim 10^{-3}$	Can be used at a variety of levels of sophistication. Good for spoting functional groups. Spectrum depends on molecular symmetry. Calculations can yield information about molecular shape and conformation of simple molecules. Useful for conformational study over wide temperature range, relatively fast time scale. Solids or solutions needed. Routine technique—also used for kinetics.
Rotational spectroscopy (microwave)	$10-10^{-1}$	Complete structural data—but only for simple molecules, gaseous sample, expensive equipment. Only compounds with dipole moment can be studied.
Nuclear magnetic resonance spectroscopy	10^{2} (in the presence of a strong magnetic field $\sim 1-3T$)	10^{-8} (up to 1 sec) Environment of and symmetry about atoms (^1H, ^{13}C, ^{19}F (^{17}O)) in molecules. Very useful for dynamic processes studied over a wide temperature range. Solution or liquid samples, expensive but routine technique.
Optical rotatory dispersion (ORD), circular dichroism (CD) See Chapter 8.		

Problems and Exercises

1. Which of the following structures represent *conformers* of **1**?

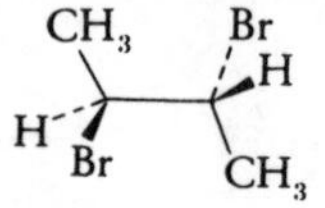

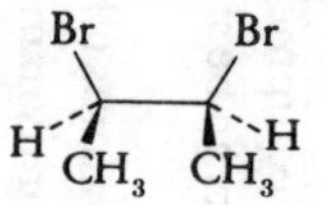

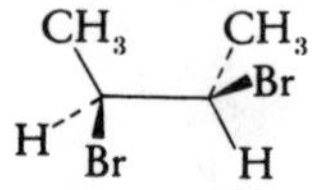

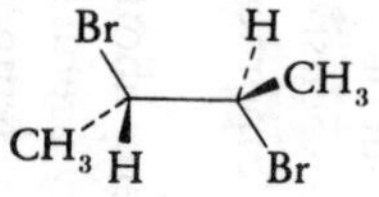

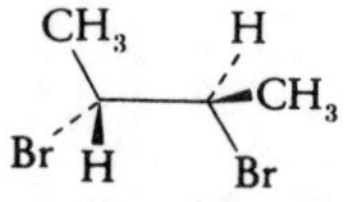

2. What is the relationship between (i) the following pairs of structures and (ii) the compounds represented by these structures?

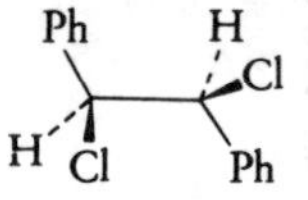

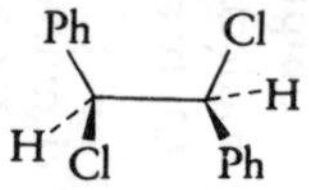

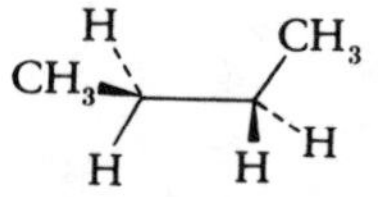

3. Differentiate, with examples, between the terms conformer, stereoisomer, atropisomer, invertomer and rotamer.

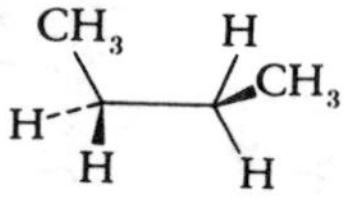

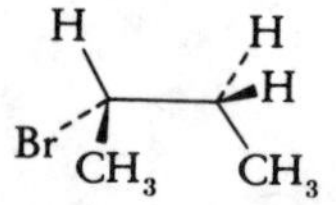

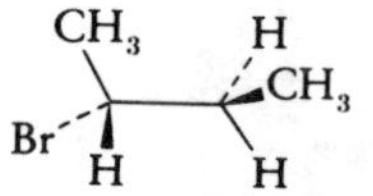

PART II

The Application of Stereochemical Principles

Whereas Section One was mostly definitions and examples this section moves into the realm of real molecules, and families of molecules.

Chapters 7 and 8 are at the very heart of organic chemistry. Cyclic systems, and cyclohexyl derivatives in particular are used time and again to exemplify one or another aspect of organic chemistry. Chiral compounds are vital to natural processes, and as such require study, they also have given insights into chemical reactivity in general.

Chapter 6 explores the stereochemistry of open-chain compounds, and the very longest open-chains, polymers are studied in Chapter 9.

CHAPTER 6

The conformations of open chain compounds

6.1 Introduction

The concept of conformation and conformational change has been introduced and defined in previous chapters. The subject of conformational analysis has importance in many areas including; the study of reaction mechanisms, the understanding of enzyme behaviour, the rationalization of polymer properties and the physical properties of compounds. Some of these are described in later chapters but first we shall examine the conformational preferences of a number of open-chain molecules, and, where possible give suitable explanations.

Before the conformations of such compounds are studied there is a brief review of the experimental basis for conformational study.

6.2 Physical methods for the determination of conformation

There are three necessary pieces of information in any conformational analysis,
- the structure of the molecules in equilibrium
- the equilibrium populations of different conformations. (Equilibrium constant, K)
- the energy barrier to interconversion of conformers. ($\triangle G^{\ddagger}$)

The major structure determination methods were summarized at the end of the previous Chapter. In the next Section we summarize how these, and other, physical methods can be used to detect conformational preferences and changes. Although we cannot review experimental results in detail it is worth remembering that almost all the results described in this book are based on structural determinations based on physical techniques.

Physical methods other than those in Table 5.1 are sometimes useful. The measurement of *dipole moments*, for example, can be revealing. There are three conformers of 1,2-dibromoethane as shown in Figure 6.1. Conformer 6.1(a) has a zero dipole moment whereas 6.1(b) and 6.1(c) have identical, finite dipole moments. A solution of 1,2-dibromoethane has a small but finite dipole moment which is lower than that calculated for 6.1(b) and 6.1(c). This shows that a mixture of conformations must be present. The dipole moment is also temperature dependent which is incompatible with completely free rotation which would give a constant, averaged value for the dipole moment.

Different species in equilibria can often be identified by a combination of Raman and infrared spectroscopies. The symmetry properties of molecules determine their Raman and ir spectra in a predictable manner; for example by comparing the two spectra, the presence or absence of a centre of symmetry in a molecule may be inferred. In the solid state the combination of Raman and infrared spectroscopies show that 1,2-dibromoethane has a centre of symmetry. Of the three conformers shown in Figure 6.1, only 6.1(a) has a centre of symmetry and therefore in the solid state the antiperiplanar conformation is the exclusive geometry for 1,2-dibromoethane. In solution, all three conformers may be observed by ir and Raman spectroscopy.

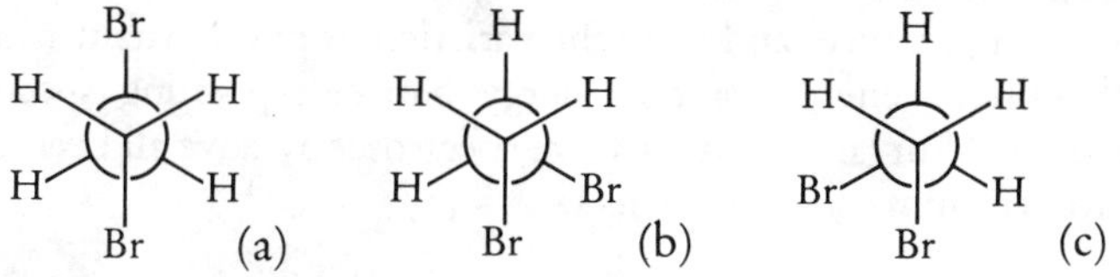

Figure 6.1   The three conformers of 1,2-dibromoethane.

The quantitative determination of equilibrium constants for conformers is also reliant on spectroscopic methods. Nmr spectroscopy, often at low temperatures, is an ideal method, but quantitative measurements by infrared or Raman spectroscopy are more difficult. The nmr spectrum of each conformer is, in principle, different. Each conformer gives rise to a number of peaks in different positions. Integration to give the areas covered by peaks gives a direct measure of their concentration. From the proportions of the various conformers, the equilibrium constant, and therefore the Gibbs free energy differences may be obtained. Most spectroscopic methods lead to calculations of Gibbs free energy differences, through measurement of equilibrium constants.

In general, an equilibrium between two (or more) species, can only be *observed* if the Gibbs free energy difference is *less* than about 15 kJ mol^{-1}. Greater free energy differences mean that the population of the higher energy species is reduced to below 1%—about the usual limit of detection.

Similarly, spectroscopic methods are often used to determine the Gibbs free energy barrier to interconversion of different species. The need for accurate measurements of energy barriers was stressed in the previous chapter. Although advanced molecular orbital calculations can be used to estimate energy barriers, the errors, when compared with experiment, are frequently large.

The basis of the spectroscopic determination of energy barriers is the difference between spectra of species that are interconverting rapidly, with respect to the time-scale of the technique, and the spectra of species interconverting relatively slowly. When the time-scale of the technique is very short compared to the rate of interconversion of conformations, a spectrum of each species is observed independently. When the rate of the process is fast compared with the time-scale of the method then the spectrum obtained is an average of the various interconverting species. At intermediate values the independent spectra change appearance and merge, or more scientifically *coalesce,* until the averaged spectrum is observed. The shapes of the spectral lines can be calculated assuming a certain rate constant for interconversion.

A series of calculated spectra are produced by a computer with a series of guessed rate constants. If an observed spectrum and a calculated spectrum are identical, then the rate constant for interconversion is that used in the interpretation of the spectrum. This process is repeated at a number of different temperatures and from the variation of rate constant with temperature the Gibbs energy of the energy barrier separating species may be determined. Nmr, ir and microwave spectroscopy have all been used extensively to determine energy barriers.

6.3 Rotations about unconjugated single bonds

In this Section the conformational behaviour of compounds containing simple, single (σ) bonds is discussed.

The simplest organic examples of conformational change, ethane, has been discussed in Section 5.3. The lowest energy conformation of ethane is staggered, of which there are three indistinguishable forms, and the energy change with dihedral angle resembles a smooth sine wave. The Gibbs free energy barrier in ethane is about 12 kJ mol^{-1} and its origin is rather obscure but often ascribed to C—H electron repulsions in the eclipsed form.

Butane, C_4H_{10} provides a slightly more complicated example of conformational analysis that brings in a second factor influencing the relative stability of conformations. The energy profile for 360° rotation about the central C_2—C_3 bond in butane is shown in Figure 6.2.

A dihedral angle of 0° is usually taken to mean that both methyl groups are eclipsed. The lowest energy conformation, and the highest energy conformation are those in which the methyl groups are as separated as possible and as close as possible, respectively (dihedral angles of 180° and 0°). This illustrates one of the *through-space* (non-bonded) interactions affecting the energies of conformations; the familiar idea of steric hindrance. In the highest energy

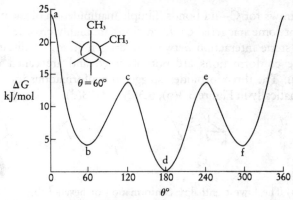

Figure 6.2 The energy profile for the rotation about the 2,3 C—C bond in butane.

conformation the methyl groups are close enough to interact to some extent. The extent of this CH_3, CH_3 interaction can be estimated at about 10 kJ mol^{-1} when compared with the second maximum where each methyl group is eclipsed by the smaller hydrogen atoms. Examination of the figure shows that even in the staggered conformations 6.2(b) and (f), the methyl groups must be interacting slightly (about 4 kJ mol^{-1}) compared with the lowest energy conformation. This interaction is often called a *gauche* interaction.

At this stage it is useful to name the various conformations available to X—CH_2—CH_2—Y molecules. The eclipsed conformation 6.2(a) is called *synperiplanar*. Conformations 6.2(b) and (f) are enantiomeric but identical in energy and probability and are called *synclinal*. The eclipsed conformations 6.2(c) and (e) are similarly enantiomeric and called *anticlinal*. Finally, the conformation with a dihedral angle of 180° is called *antiperiplanar*. All conformations intermediate between energy minima and maxima are simply called *skewed* conformations.

Longer chain alkanes may be treated similarly and for all C_nH_{2n+2} hydrocarbons the lowest *enthalpy* conformation is always the all-antiperiplanar conformation. The lowest enthalpy conformation of hexane is shown schematically in Figure 6.3(a). You should make a model of this to confirm that all the hydrogen atoms are eclipsed and each carbon atom is antiperiplanar to all others. As hydrocarbon chains become longer it is often observed that there is only a very, very small proportion of the lowest enthalpy conformer present in solution, or the gas phase. One reason for this is that the *entropy* term favours more randomized conformations. Even for such a small chain as that in hexane the all-antiperiplanar conformation accounts for only 33% of the conformations at 300 K, reducing to 16% at 600 K as the $T\Delta S$ term, in $\Delta G^\ominus = \Delta H^\ominus - T\Delta S^\ominus$, becomes more important at the higher temperatures.

The steric effect on conformational preference is considerable in hydrocarbon chains. Make a model of hexane, with 5 cm straws for C—C bonds

69

and 3.5 cm straws for C—H bonds. Simple manipulation of the model will show you that some staggered conformations are highly crowded as there is considerable steric interaction between hydrogen atoms on distant carbon atoms. These conformations are not observed and are called *forbidden conformations*. The three forbidden staggered conformations for hexane are shown schematically in Figure 6.3(b), 6.3(c) and 6.3(d).

(a)   (b)   (c)   (d)

Figure 6.3   (a) The lowest enthalpy conformation of hexane; (b, c, d) forbidden conformations of hexane.

For small hydrocarbon chains, up to about $n = 12$, the *exclusive* form in the solid phase is all staggered. As there are only small energy differences between conformations the energy gained by more efficient packing of the regular, all antiperiplanar chains more than offsets the statistical probability of finding the more randomized forms. The packing of longer chain hydrocarbons such as polymers is explored in a later chapter.

So far, the effects on conformer distribution that have been discussed are the through-space, steric effect which varies with dihedral angle; the little-understood electronic effect that gives preference to staggered forms in, say, ethane; and the statistical or entropy effect that becomes important when there are several forms with the same energy and probability.

There are other, through-space, interactions that can dominate in open-chain non-conjugated compounds. One of these is *intramolecular hydrogen-bonding*. There are three staggered conformations available to 2-chloro-ethanol, $ClCH_2CH_2OH$, as shown in Figure 6.4. The two synclinal conformations 6.4(b) and 6.4(c) predominate, in both the solution and gas phase, over the sterically-favoured, antiperiplanar form 6.4(a). In the anticlinal conformations the chlorine atom and the O-bonded hydrogen atom are sufficiently close to enable significant hydrogen bonding, as illustrated by the hatched lines in the figures. This hydrogen bonding lowers the free energy of the synclinal forms by about 10 kJ mol⁻¹ so that they are favoured by about 4 kJ mol⁻¹. Fluorine, being more electronegative than chlorine, forms stronger hydrogen bonds so that 2-fluoroethanol is more favoured in the synclinal conformation relative to the antiperiplanar conformer by about 8 kJ mol⁻¹.

Electrostatic attraction is also responsible for favouring synclinal conformations of apparently sterically hindered molecules. Acetylcholine, $CH_3COOCH_2CH_2\overset{+}{N}(CH_3)_3$, is found in the nervous system and plays a crucial role as a neurotransmitter (i.e. sending messages round the nervous system). As a prelude to the understanding of the action of acetylcholine (and many other natural chemicals and drugs) an understanding of its conformation in solution is crucial. Again the synclinal conformations are favoured, (Figure 6.5); in this example because electrostatic attraction between the positively charged quatenary ammonium nitrogen atom and the slightly negative oxygen atoms in the acetyl group dominates the conformation in solution. In fact the electrostatic attraction is so strong in this example that the dihedral angle of the most favoured conformation is probably closer to 0° than to 60°.

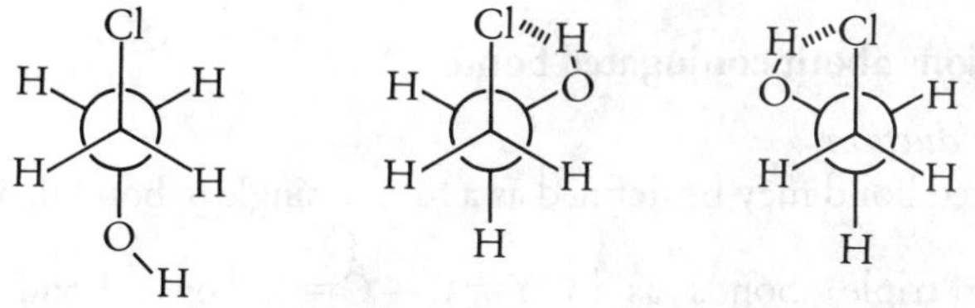

Figure 6.4 (a) The antiperiplanar conformation of 2-chloroethanol; (b and c) the favoured synclinal conformation, stabilized by intramolecular hydrogen bonds.

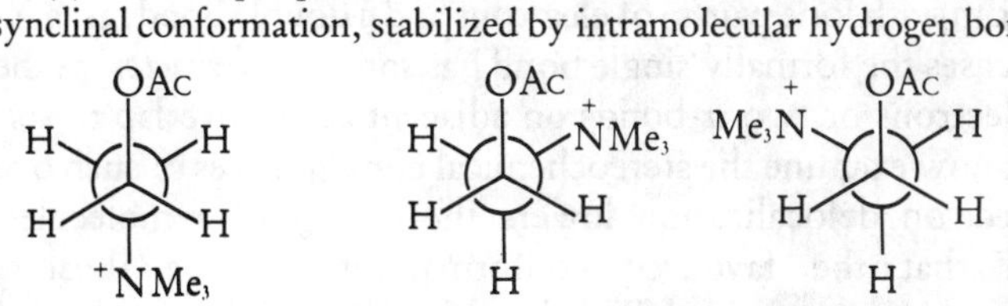

Figure 6.5 Conformations of acetylcholine. The synclinal conformations are favoured by electrostatic attractions.

For the final examples of the conformational preference of compounds with unconjugated single bonds we shall examine briefly compounds of the type $R_3C{-}\overset{|}{C}{=}X$, where $X = O$, N, or C. It has been discovered that free energy barriers to rotation in these systems are almost all rather lower than $C{-}C$ barriers and fall in the range of 3–8 kJ mol^{-1}. In almost all cases the preferred conformation is one with the double bond eclipsed as shown in Figure 6.6. This is sometimes stated as 'preferred eclipsing of the double bond'. The reasons for this preference are, once again, rather obscure but may be connected with a stabilizing interaction between the R—C σ-bonds and the C = X π-electrons. This type of interaction, called *hyperconjugation* or σ–π vertical stabilization, is rather controversial for such systems and best left to more advanced texts where a detailed knowledge of molecular orbital theory may be assumed.

The only common exceptions to the rule of preferential double bond eclipsing occur when fluorine is incorporated into the molecule. Fluoropropanone exists in a conformation with a fluorine-oxygen dihedral angle of **180° as shown in Figure 6.6(b) on p. 72. It is probable that dipole-dipole repulsion between the slightly negatively charged oxygen and fluorine atoms forces this conformation on the molecule.**

71

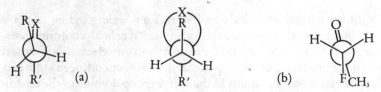

Figure 6.6 (a) Newman projections of the preferentially eclipsed double bond. (b) The preferred conformation of 2-fluoropropanone. The fluorine and carbonyl are antiperiplanar, possibly due to dipole–dipole repulsions.

6.4 Rotations about conjugated bonds

6.4.1 Introduction

A conjugated bond may be defined as a formal single σ-bond flanked by two double (or triple) bonds as in $Y{=}C{-}\overset{|}{C}{=}X$, or a bond between a heteroatom with lone pair(s) of electrons and a double bond, as in $\ddot{X}{-}\overset{|}{C}{=}Y$. In both cases the formally single bond has some 'π-*character*' as the lone pair and π-electrons or two π-bonds on adjacent atoms overlap to some extent. We shall now examine the stereochemical consequences of such overlap.

As electron delocalization lowers the energy of a molecule it can be assumed that the favoured conformations will be those in which delocalization is maximized. This is achieved when the two formal π-bonds are coplanar, or when the heteroatom lone-pair of electrons is coplanar with the π-bond. It also follows that if the delocalized conformations are stabilized, a barrier to rotation *greater* than that in simple σ-bonds may be expected. With these simple ideas in mind we can survey the conjugated systems by class. It is preferable to review them in this way as each class has its own distinctive structural features.

6.4.2 Dienes, conjugated aldehydes and ketones

These systems ${>}C{=}C{-}\overset{|}{C}{=}X$ are all relatively weakly delocalized, although there is sufficient delocalization to ensure that the C, C, C X framework is planar in the most stable forms.

Butadiene is capable of existing in two planar forms called the *s-cis* and *s-trans* forms, (Figures 6.7(a) and 6.7(b) respectively). The designation *s* indicates that the disposition about the formal single bond is being described and *cis* is equivalent to a synperiplanar arrangement of double bonds and *trans* similarly to an antiperiplanar conformation. The *E, Z* nomenclature is more useful (and is unambiguous); 6.7(a) is *s-Z* and 6.7(b) *s-E*. The *s-E* conformer is favoured by about 10 kJ Mol^{-1} owing to some steric hindrance in the *s-Z* form. In fact the *s-Z* form is not perfectly planar but has a dihedral angle of about 15°. The free energy barrier to rotation in butadiene (interconversion of the *s-Z* and *s-E* forms) is 30 kJ mol^{-1} which is about 18 kJ mol^{-1} greater than that in ethane; this extra barrier being almost all ascribable to the lowering of the ground state energy by conjugation.

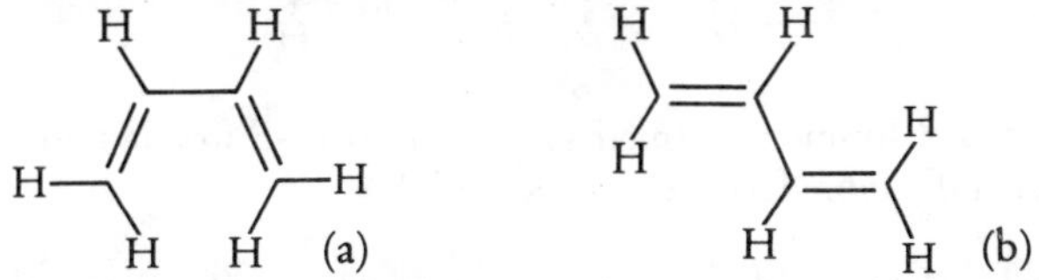

Figure 6.7 (a) s-Z-butadiene; (b) s-E-butadiene. The s-E conformer is the more stable.

Propenal (acrolein, $CH_2 = CHCHO$) is similar in conformation to butadiene; the s-E conformer is favoured by 8.7 kJ mol^{-1} and the barrier to rotation is 21 kJ mol^{-1}. Conjugation is less effective in propenal as the electron density in the C=O bond is concentrated nearer the oxygen atom, resulting in a smaller overlap between the C=C and C=O π-bonds. Substitution of a methyl group for the 2-hydrogen atom (to give 2-methylpropenal) increases the rotational free energy barrier to 23 kJ mol^{-1}, through the slightly greater steric effect.

Conjugation with aromatic rings is also observable, as in benzaldehyde, C_6H_5CHO, where the barrier to C—C rotation is 20 kJ mol^{-1}, very similar to that in propenal.

6.4.3 Esters, acids, amides and thioamides

These molecules all contain heteroatoms, O, N or S conjugated with a carbonyl or thiocarbonyl π-bond. They are all planar, for simple systems, with two possible conformers, often called s-cis or s-trans although again the E, Z nomenclature is less ambiguous (Figure 6.8).

Figure 6.8 The two conformers of esters, amides, etc.

The conformational preferences are controlled by a combination of steric and electronic effects, although many authorities believe steric effects to dominate. The free energy barrier to rotation is strongly dependent on the extent of conjugation but very large groups can also affect the barrier substantially.

Esters and acids are the least delocalized of this group which is reflected in their relatively low barriers to rotation about the C—OR bond of about 40 to 50 kJ mol^{-1}.

Formic acid shows a pronounced preference for the Z conformer (Figure 6.9(a)) which is some 8 kJ mol^{-1} more stable than the E conformer. This prediliction is similar to that displayed by ketones and aldehydes with preferential eclipsing of the double bond. The free energy barrier to interconversion in formic acid is 46 kJ mol^{-1}.

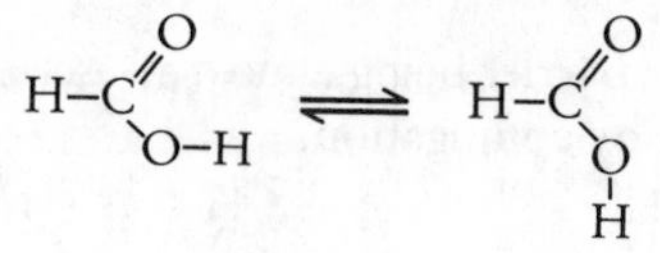

Figure 6.9 (a) *s-Z* formic acid (methanoic acid); (b) *s-E* formic acid. The eclipsed double bond in (a) is more favoured by 8 kJ mol^{-1}.

Substitution of the acidic proton by an ethyl group to give ethyl formate still results in the Z isomer predominating in this example by about 11 kJ mol^{-1}. The barrier to rotation is again about 40 kJ mol^{-1}.

Formates with particularly bulky groups, such as tertiary butyl have rather more E isomer present, the extent of which depends on the solvent polarity; the Z isomer predominating in non-polar solvents.

Ethanoates, and other aliphatic esters RCOOR' have an even greater tendency to form Z conformers; a suitable explanation being that steric hindrance between R and R' is increased in the E isomer as the size of the group attached to the carbonyl carbon increases.

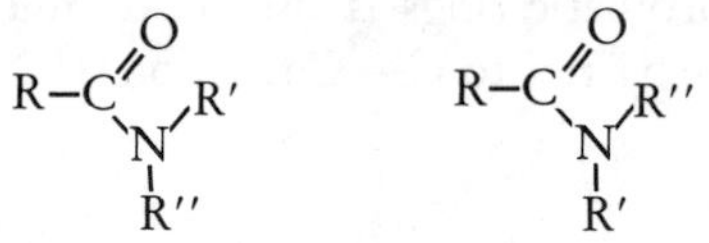

Figure 6.10 The two amide conformers, *s-Z* (a) and *s-E* (b), when R' has preference over R'' and O is preferred to R.

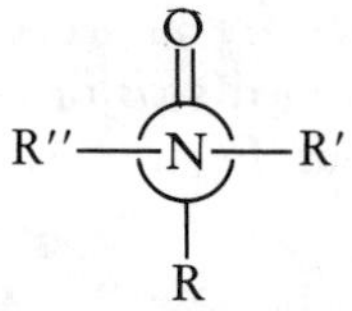

Figure 6.11 The geometry of the activated complex for amide C—N bond rotation.

The conjugation in amides is quite marked and reflected in the much higher barriers to rotation than in acids and esters ($\sim$ 60–90 kJ mol^{-1} cf. $\sim$ 40–50 kJ mol^{-1}). The stable rotational forms are equivalent to those in other conjugated systems. The two rotamers of amides RCONR'R'' are shown in Figure 6.10 together with their designation. Deconvoluting the various effects on conformer distribution and ratio is difficult as there are the steric *and* electronic effects of R, R' and R'' to consider. Certain crude generalizations can be made on the evidence obtained from, probably, hundreds of studies on amide conformations. First, the barrier height to E, Z interconversion is correlated with the size of R' and R''. As the size of R' and R'' increases, the ground states, the conformers 6.10(a) and 6.10(b), become increasingly crowded and consequently their Gibbs free energy increases. The energy of the transition state is relatively unperturbed by steric effects as illustrated by the Newman projection in Figure 6.11. The net result is that the Gibbs free energy barrier to rotation is lower for systems with large substituents. Examples of energy barriers to rotation of amides are shown in Table 6.1; this is own on pp. 76–77.

Electronic effects are also important. Substituents that tend to decrease the extent of amide π-bonding tend to lower the barrier to rotation. The C—N π-electron density can be lowered by competing conjugation by R, as with the vinyl group for example. Similarly the phenyl group conjugates with the carbonyl group, but steric effects may come into play here.

Substituents, R' and R'' that increase the electron density on the nitrogen atom should increase the conjugation and therefore also increase the rotation barrier.

Thioamides, $RCSNR_2'$ also show restricted rotation about the C—N bond with rotation barriers generally about 10 kJ mol^{-1} higher than the corresponding amides. Similarly enamines C=C—NR$_2$ are also conjugated with restricted C—N rotation, although in this case conjugation is not strong, with a correspondingly lower barrier to rotation, about 10–15 kJ mol^{-1} *less* than in similar amides.

6.4.4 Biphenyls

Although there is conjugation between the phenyl rings in planar C_6H_5— C_6H_5 the most favoured conformation in solution is clinal (i.e. the benzene rings are not in the same plane) with a dihedral angle of about 20°. Steric repulsion between the *ortho* (2-) hydrogen atoms is sufficient to offset the energy lost on conjugation and make the molecular non-planar. The barrier to rotation, and the dihedral angle between the rings, can be increased dramatically by substituting the *ortho* hydrogen atoms by larger groups. The barrier to rotation in biphenyl-2,2'-disulphonic acid (Figure 6.12) is greater than 100 kJ mol^{-1} and this acid has been resolved into the two enantiomeric forms 6.12(a) and 6.12(b). Models should convince you that these structures represent enantiomers. The barrier to rotation is strongly influenced by the steric effect of the substituents, it is reduced to 57.4 kJ mol^{-1} when the two *ortho* substituents are methoxy groups (CH$_3$O). A list of typical barriers to rotation in biphenyl derivatives is found in Table 6.1. In general biphenyls substituted at all four *ortho* positions (2,2',6,6' substitution) are the most crowded and consequently have the greatest barrier to rotation. It should be noted the highest energy point on the biphenyl rotation profile is the conjugated, planar conformation, in which steric hindrance is greatest.

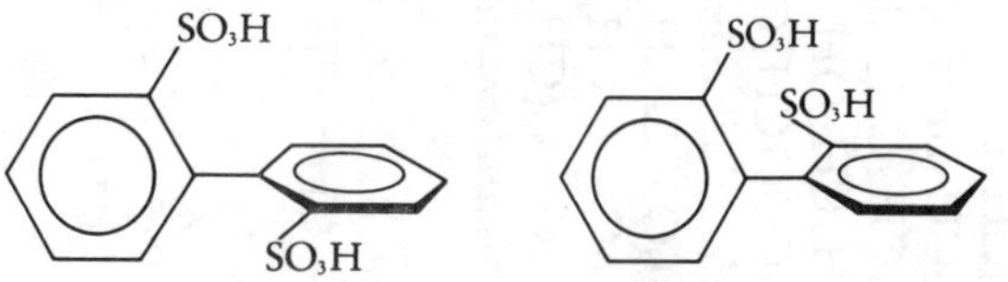

Figure 6.12 Enantiomeric forms of biphenyl-2,2'-disulphonic acid, that can exist as pure compounds.

Table 6.1. Barriers to rotation and conformational preferences of some formally single bonds

Compound	$\Delta G^{\ddagger}$/kJ mol^{-1}	Conformational preference
Simple alkanes XYZC—CX'Y'Z'		
CH$_3$CH$_3$	12–20	staggered
CH$_3$CH$_2$F	12	staggered
CH$_2$CH$_2$Br	13.9	staggered
CH$_3$CH$_2$CH$_2$CH$_3$	15.4	antiperiplanar
ClCH$_2$CH$_2$OH	24	synclinal
CH$_3$COOCH$_2$CH$_2$$\overset{+}{\text{N}}$(CH$_3$)$_3$		synclinal
Hindered alkanes		
(CH$_3$)$_3$C—C(CH$_3$)$_2$H	29.2	
(CH$_3$)$_3$C—C(CH$_3$)$_2$F	34	
(CH$_3$)$_3$C—C(CH$_3$)$_2$I	46.6	
Alcohols and amines R$_3$C—OH, R$_3$C—NH$_2$	5–10	staggered
Alkenes R$_3$C—C=X		
CH$_3$—CH=CH$_2$	3–8	double bond eclipsed
	8	double bond eclipsed
(structure: H, H / CH$_3$, CH$_3$ with C=C)	3	
Aldehydes and ketones R$_3$C—CX (with C=O)		
(CH$_3$)$_2$C=O	2–6	double bond eclipsed
CH$_3$—C(O)Cl	3.4	double bond eclipsed
FCH$_2$C(O)CH$_3$	4.2	CO and F antiperiplanar

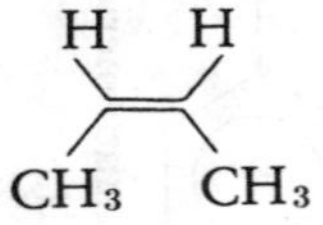

Conjugated systems

dienes $>C{=}C{-}C{=}C<$	30–40	*s-E*
$H_2C{=}CH{-}CH{=}CH_2$	30	*s-E*
enones $>C{=}CH{-}CHO$	20–30	*s-E*
$CH_2{=}CH{-}CHO$	21	*s-E*
$C_6H_5{-}CHO$	20	planar
Esters and acids		
$RC(O){-}OR'$, $RC(O)OH$	40–50	*s-E*
$HC(O){-}OH$	46	*s-E*
$HC(O){-}OC_2H_5$	40	*s-E*
Amides $RCONR'R''$	60–90	*s-E*
$HC(O)NH_2$	74.4	planar
$HC(O){-}N(CH_3)_2$	88	planar
$CH_3C(O){-}NH_2$	71	planar
$CH_3C(O){-}N(CH_3)_2$	76	planar
$HC(O){-}NHCH_3$		Z
$C_2H_5C(O){-}NHC_2H_5$		Z
$HC(O){-}NHC(CH_3)_3$		E
Enamines $R_2C{=}CH{-}NR'_2$	50–80	Z
Thioamides $RC(S){-}NR'_2$	70–100	Z
Biphenyls	40–120	clinal
$o{-}CH_3OC_6H_5{-}C_6H_5OCH_3{-}o$	57.4	clinal
$o{-}CH_3C_6H_5{-}C_6H_5CH_3{-}o$	73	clinal
$o{-}C_2H_5OC_6H_5{-}C_6H_4(COOH)(NO_2){-}o$	84	clinal
$o,o{-}(CH_3O)(CH_3COO)C_6H_4{-}C_6H_4(OCH_3)(OOCCH_3){-}o,o$	105	clinal (resolvable into enantiomers)
$o{-}H_2NC_6H_5{-}C_6H_5NH_2{-}o$	>100	clinal (resolvable into enantiomers)
$o{-}HO_3SC_6H_5{-}C_6H_5SO_3H{-}o$	>100	(resolvable into enantiomers)

6.5 Summary of Sections 6.3–6.5

1. There is an 'intrinsic' Gibbs free energy barrier to rotation about $sp^3C—Csp^3$ bonds of rather uncertain origin, and of about 12 kJ mol^{-1}. Most simple C—C single bond rotations have barriers of 12 to 20 kJ mol^{-1}.

2. For simple hydrocarbons the lowest energy conformation is that with the bulky groups as far removed as possible. This is the antiperiplanar form for XCH_2CH_2Y. Steric hindrance, a through-space interaction is responsible for this preference.

3. For straight-chain alkanes the lowest *enthalpy* conformation is always the all-antiperiplanar form in which all carbon atoms are antiperiplanar to their respective neighbours.

4. In solution and in the gas phase, the amount of the all antiperiplanar form of alkanes decreases with chain length and temperature, as the entropy term starts to dominate.

5. For long chain alkanes some staggered conformations are 'forbidden' as they require two or more atoms to occupy the same space.

6. Intramolecular hydrogen bonding between neighbouring substituents, and electrostatic attraction can stabilize synclinal conformations, or even synperiplanar conformations at the expense of antiperiplanar conformations.

7. The barriers to $sp^2C—sp^3C$ rotation are generally lower than $sp^3C—sp^3C$ barriers and are about 3 to 8 kJ mol^{-1}.

8. The double bond in such systems as $R_3C—C=X$ is usually eclipsed by a substituent R, sometimes expressed as preferential eclipsing of the double bond.

9. Dipole-dipole repulsions as in $FCH_2C(O)CH_3$ can lead to antiperiplanar conformations being preferred.

10. Conjugation usually increases rotation barriers.

11. Even weakly conjugated systems such as dienes have rotation barriers of about 30 to 40 kJ mol^{-1}.

12. For dienes and analogues there are two conjugated conformers, called either *s-Z(s-cis)* or *s-E(s-trans)* according to their arrangement of double bonds. For steric reasons the *s-E* conformer is favoured over the *s-Z* conformer.

13. Conjugation is greater in esters, acids and amides than in dienes and enones. Esters are least conjugated with a barrier to rotation about the C—OR bond of 40 to 50 kJ mol^{-1}.

14. The double bond in amides, esters, acids and thioamides is again preferentially eclipsed (Z-conformation dominates).

15. Amide barriers to C—N rotation fall in the region 60–90 kJ mol^{-1} and are affected by steric effects (bulky groups lower the barrier) and electronic effects (increasing C—N π-character increases the barrier).

16. Biphenyls, although conjugated have crowded planar conformations

owing to interactions between *ortho* hydrogen atoms. Substitution of *ortho* hydrogens by bulkier groups dramatically increases the energy barrier to rotation to such a degree that resolution of some $o—C_6H_5X—C_6H_5X—o$ compounds into enantiomers is possible.

Problems and Exercises

1. Use models to help you to draw all of the staggered conformations of pentane. Which of these is the forbidden conformation? By analogy with butane, suggest which of the allowed conformations are of the highest and lowest enthalpies.

2. The energy profile for the rotation about the 2,3 C—C bond in R-2-iodobutane is shown in Figure 6.13. Assume that the energies of the confirmations are determined by steric effects alone. (van der Waals radii are given in Chapter 1.) Draw Newman projections of the conformations corresponding to positions a–d on the figure. Why is the plot of energy against dihedral angle unsymmetrical?

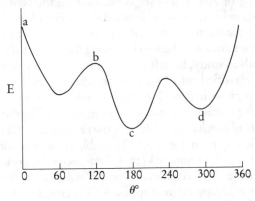

Figure 6.13   The energy profile for the rotation about the 2,3 C—C bond in R-2-iodobutane.

3. Draw the preferred conformation(s) of
   (a) $ClCH_2CH_2Br$
   (b) $CH_3COOCH_2CH_2\overset{+}{S}(CH_3)_2$
   (c) FCH_2CH_2OH
   (d) $E—H_2C=CH—CH=CH—CH=CH_2$
   (e) CH_3CH_2CHO
   (f) CH_3COOCH_3
   (g) $HCOCH_2F$
   (h) $HCON(CH_3)_2$
   (i) $o—CH_3C_6H_4—C_6H_4CH_3—o$
   Where possible, give brief explanations for your choice of conformations.

CHAPTER 7

The conformations of saturated cyclic systems

7.1 Introduction

The presence of a ring of atoms imposes constraints on the conformational freedom of molecules which, in turn, limits the number of available conformers. This by no means limits the interest in cyclic molecules; on the contrary many chemists consider them to be among the most challenging and interesting compounds to study, both for their conformations and reactivity.

Rings are usually classified into four categories; *small rings,* with 3 or 4 members; *normal rings* with 5 to 7 members; *medium rings* with 8–11 members and *large rings* with more than eleven ring atoms. The classification originates in the *heats of combustion* $\triangle H^{\ominus}$ for the carbocycles $(CH_2)_n$ which are given in a modified form in Table 7.1. The table gives two numbers for each molecule. The first is the *strain,* in kJ mol^{-1}, for each CH_2 group, relative to a straight chain alkane. This is a measure of the 'stability' of the molecule and shows the excess enthalpy content for *each* CH_2 group in the molecule compared with its open-chain analogue. As an example, cyclopropane $(CH_2)_3$ has an excess enthalpy of 38.6 kJ mol^{-1} for each CH_2 group, making that molecule highly strained. The second figure in the table is the total strain for each molecule, again relative to its open-chain analogue; 116 kJ mol^{-1} for cyclopropane.

Examination of the table shows that small rings are highly strained, normal rings show a relatively lower strain, medium rings are increasingly strained and large rings are strainless.

The two major contributors to ring strains are called *Baeyer strain* and *Pitzer strain*. Both effects are analogous to strains in open-chain compounds.

Baeyer strain, or *angle strain* is a consequence of deforming bond angles from their optimal tetrahedral value of about 109°. Pitzer strain arises from non-bonded (predominately steric) interactions.

Any ring can suffer from a combination of Baeyer and Pitzer strains although the emphasis is different for each ring class. For example, the predominant strain in small rings is Baeyer strain, especially in cyclopropane where the internuclear bond angles can only be 60°.

Table 7.1. Strain energy per methylene group in cycloalkanes

Carbocycle $(CH_2)_n$	n	strain/CH_2/kJ mol^{-1}	strain/molecule/ kJ mol^{-1}
small rings	3	38.6	116
	4	27.2	109
'normal' rings	5	5.0	25.1
	6	0.0	0.0
	7	3.8	26.4
medium rings	8	5.0	40.2
	9	5.9	52.8
	10	5.4	54.5
	11	4.2	46.0
large rings	12	1.2	15.0
	13	1.7	21.7
	14	0.0	0.0
	16	0.4	6.7
	17	−0.8	−13.6
	∞	0.0	0.0

The original angle strain theory, postulated in 1885 by Adolph von Baeyer assumed that all rings were *planar*. It was therefore difficult to understand the stability of the normal rings (and particularly cyclohexane where bond angles should be 120° for a planar molecule). As early as 1890 H. Sachse suggested that rings could adopt non-planar conformations to overcome Baeyer strain, but there was a great deal of contemporary hostility to his ideas. E. Mohr revived the idea in 1918 but it was not until the 1950's that conformational preferences in rings became accepted. Then, O. Hassel determined the structures of cyclohexanes by X-ray analysis and D.H.R. Barton carried out conformational analyses on ring systems (steroids in particular) and they were jointly awarded the 1969 Nobel Prize for chemistry for their seminal work.

Now, the structures of many ring systems have been studied and the non-planarity of most aliphatic rings is well-established. Once non-planar ring conformations were accepted as fact, it became obvious that rings will adjust their conformations to attain a minimum energy so that strain is spread optimally between the Baeyer and Pitzer contributions.

In the normal rings, both Baeyer and Pitzer strain can be minimized by the adoption of the non-planar conformations described in subsequent Sections of this chapter.

Medium rings can adopt conformations in which Baeyer strain is practically eliminated but Pitzer strain is always present. One particular form of Pitzer strain may be significant for medium rings, and that is *transannular interaction* where two hydrogen atoms may be competing for the same region of space within a ring.

Large rings, as demonstrated by their strain energies, are sufficiently flexible to adopt conformations that are equivalent in energy to their open-chain counterparts. They have no Baeyer or Pitzer strain more significant than in a straight-chain alkane.

When discussing the stability of aliphatic rings (alicycles), which is a *thermodynamic* property, there is often confusion with their *ease of formation*, which is of *kinetic* origin.

The easiest rings to form are three membered—but these are also the easiest rings to break. Three membered rings form readily because any three adjacent atoms are necessarily in a plane and have a geometry appropriate for cyclization.

Four membered rings are relatively difficult to form, not simply because of their inherent strain (*cf.* 3 membered rings) but more because of the need to cyclize four atoms from the high-energy synperiplanar conformation. This constraint is less important for five and six membered rings. As rings get larger the conformational energies are not so important as the entropy effect, since the population of the appropriate conformation has an increasingly low probability. The ease of formation of seven and greater membered rings is severely reduced by the entropy effect.

The remainder of this chapter is concerned only with the stereochemistry of ring systems, with a special emphasis on conformational analysis. It is simplest to start this study with the six membered rings as much of the understanding of alicyclic compounds depends on a familiarity with cyclohexane and its derivatives.

7.1.1 Summary of Section 7.1

1. Alicyclic rings are classified as: small, 3–4 members; normal, 5–7 members; medium, 8–11 members; large 7–11 members.
2. Ring strain, as measured by heat of combustion is large for small rings, low for normal rings, higher for medium rings and absent in large rings.
3. The cause of strain in rings is a combination of Baeyer strain (angle deformation from 109°) and Pitzer strain (non-bonded interactions).
4. Rings adopt non-planar conformations to distribute strain optimally between Baeyer and Pitzer strain.
5. The ease of ring formation is *not* related to the strain (internal energy) of

82

the rings, but to kinetic factors depending on the transition state conformational energy and probability. The ease of formation of rings is $3 \gg 4 < 5 \leq 6 > 7 \geq 8$.

7.2 The Normal Rings: Cyclohexane and its derivatives

7.2.1 Cyclohexane

The frequent use of molecular models is essential for a thorough understanding of cyclohexane and its derivatives. In fact it was primarily through the study of specially commissioned models (from a watchmaker!) that Sir Derek Barton was able to formulate his theories on the preferred conformation of six membered rings. Cyclohexane, C_6H_{12}, exists predominantly in a non-planar, puckered conformation called the *chair* conformation, that is illustrated in Figure 7.1. It is advisable to make a model of chair cyclohexane during study of this section. If your initial model is flexible in feel, hold it by two opposing carbon atoms (say C-1 and C-4 and alter the conformation until a structure looking like that in Figure 7.1 is obtained. A model of a chair conformation of cyclohexane has a rigid feel.

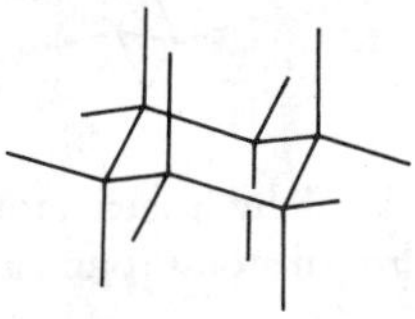

Figure 7.1   The favoured chair conformation of cyclohexane.

The chair conformation is at the minimum energy possible for C_6H_{12}. A Newman projection along two pairs of parallel C—C bonds (say 1,2 and 4,5 or equivalently 2,3 and 5,6) as shown in Figure 7.2 immediately reveals the important stereochemical features of a chair cyclohexane conformation. All bonds are staggered so that Pitzer strain is minimized. The only steric strain arises from gauche butane-like interactions between neighbouring methylene groups. Baeyer strain is non-existent as all bond-angles can be tetrahedral. The C—C—C bond angles are not exactly 109.5° but at 111.1° are almost identical to those in straight chain alkanes. The dihedral angles, θ, are not exactly 60° but closer to 55° causing a very slight flattening of the ring. All this helps to explain why the chair has the lowest possible conformation energy for any cyclohexane conformation, as each interaction is at a minimum.

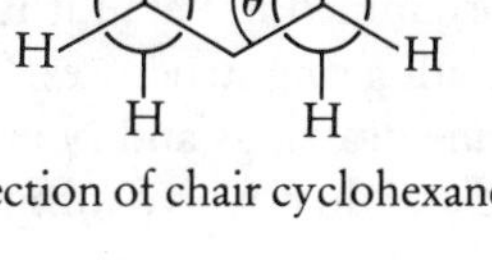

Figure 7.2   A Newman projection of chair cyclohexane.

A further stereochemical feature of chair cyclohexane is also apparent from both Figures 7.1 and 7.2 and that is the existence of *two* types of hydrogen atoms. The *ring plane* in chair cyclohexane is defined as an imaginary plane running through the molecule so that all carbon atoms are equidistant from that plane. One type of hydrogen atom is perpendicular to the ring plane, alternating above and below the plane round the ring. These hydrogen atoms are called *axial*, and may be identified by placing a model of chair cyclohexane on a smooth horizontal surface. The model will stand firmly on three axial hydrogen atoms and the other three axial hydrogen atoms will point vertically upwards.

The other six hydrogen atoms are called *equatorial* and run roughly parallel to the ring plane, but again disposed alternately above and below the plane. Each type of hydrogen atom is illustrated separately in Figure 7.3. All six axial atoms are indistinguishable from each other and similarly for the equatorial atoms. However, equatorial hydrogen atoms are readily distinguished from axial hydrogen atoms in the model.

(a) (b)

Figure 7.3 (a) Alternation of axial hydrogen atoms round a cyclohexane ring. (b) Alternation of equatorial hydrogen atoms round a cyclohexane ring.

One of the reasons why the non-planar forms of cyclohexane did not become acceptable until relatively recently was that there should be chemical consequences of having two distinct sets of hydrogen atoms. Given the analytical techniques of the early and mid twentieth century no distinction could be made between axial and equatorial hydrogen atoms, and only one monosubstituted cyclohexane seemed to exist for any substituent.

It is now known that cyclohexane appears to have one equivalent set of hydrogen atoms because *one chair form is in rapid equilibrium with another indistinguishable chair*. During the ring exchange the axial and equatorial hydrogen atoms are interchanged. *This is a conformational change that requires no bond breaking or making*. The chair–chair interconversion is easily illustrated by models. Label the axial *or* equatorial atoms with different coloured straws or by adding a coloured centre to one set. Now hold the ring by opposing carbon atoms and simultaneously twist each towards the centre of the ring, crossing the ring plane. Another chair is formed with axial-equatorial positional interchange. This process is sometimes called a ring flip.

Before this exchange is examined in detail it is necessary to present some different methods for illustrating ring structures. So far we have illustrated the cyclohexane structure by line drawings and Newman projections. These are usually somewhat idealized and do not give sufficient quantitative data on the structures for conformational analysis. In the formulation given in Figure 7.4(a) each ring C—C bond is given a number that corresponds to the torsion angle, θ, and a sign corresponding to a clockwise $(+)$ or anticlockwise $(-)$ rotation for θ. The convention used is the same as that in the previous chapter. In Figure 7.2, θ is positive. An alternative, non-quantitative presentation is shown in Figure 7.4(b), where each carbon atom is labelled as positive or negative depending on whether it is respectively, above or below the ring plane. The ring plane is defined by the dashed lines that lie in that plane.

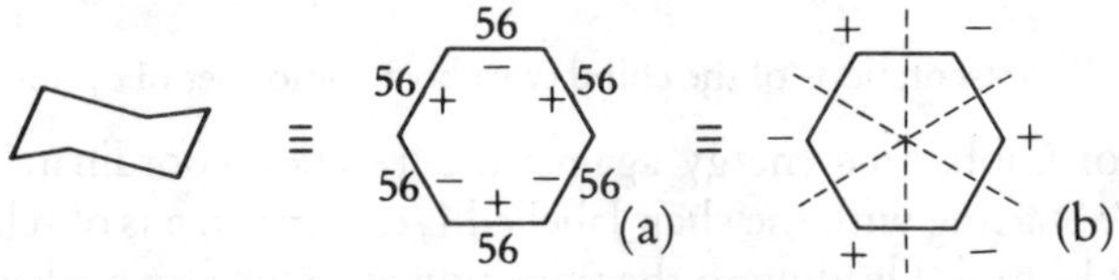

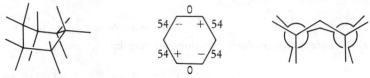

Figure 7.4   Alternative representations of chair cyclohexane.

A combination of structural representations will now be used to help explain the conformational changes of the cyclohexane ring during chair–chair interconversions. The rigid chair conformers are energy minima but between these there are a number of *flexible* forms with higher energies. As the flexible conformations have a greater conformational freedom they also have higher entropies. It is useful to examine the structures of some of the flexible conformations before looking at the detailed changes during ring flips. Chair cyclohexane has a well-defined structure as it is rigid, but the flexible forms are, by their nature, less well-defined. We shall concentrate on structures that lie at energy maxima or minima.

The easiest flexible cyclohexane structure to visualize is the *boat* conformation shown in Figure 7.5. The boat conformation lies at a local energy maximum as two bonds, labelled 0, 0, are completely eclipsed, thereby increasing the Pitzer strain. The symmetry of the boat is reduced relative to the chair. The chair has point group $\mathbf{D}_{3d}$ and the boat $\mathbf{C}_{2v}$. In the boat two of the six carbon atoms have distinctly different environments from the other four.

Figure 7.5   Representations of boat cyclohexane.

Apart from the chair, the only other species at an energy minimum is the *twist boat* conformer, which is of higher energy than the chair. Figure 7.6 (p. 86) give various representations of the twist boat. These figures show that Pitzer strain is relieved relative to the boat as no bonds are eclipsed, and the smallest torsion angle is 31°. The symmetry of the twist boat is again reduced; it has point group $\mathbf{D}_2$ and is chiral. Both the chair and boat cyclohexanes can exist in a number of indistinguishable forms but the twist boat can exist in two distinguishable, enantiomeric sets.

The chair, boat and twist boat all feature in the conformational changes during chair–chair interconversions of cyclohexane.

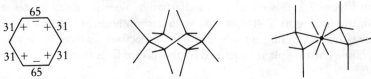

Figure 7.6   Representations of the chiral twist boat conformer of cyclohexane.

A plot of Gibbs free energy against the 'reaction coordinate' is given in Figure 7.7. Starting with the chair labelled 1, the first step is of relatively high energy, 42 kJ mol^{-1} leading to the transition state, or activated complex, for the process. Two structures appear to be possible for the activated complex. The one given on Figure 7.7 is of C_2 symmetry with four coplanar carbon atoms and it leads directly to a twist boat conformation. Although activated complexes cannot be studied directly, calculations and chemical intuition suggest that the C_2 transition state is of slightly lower energy than its competitor, the C_s envelope shown in Figure 7.8. The envelope activated complex leads directly to a boat conformation.

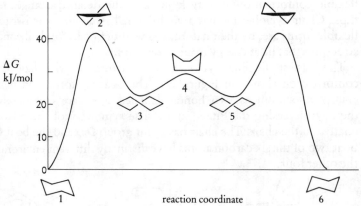

Figure 7.7   Energy changes during a cyclohexane ring flip.

Figure 7.8   The envelope conformation of cyclohexane.

The first step, whatever the transition state, is of higher energy than other steps because energy is needed to accommodate the necessary bond angle changes which lead to a higher Baeyer strain. In general conformational changes requiring bond angle deformations are called *inversions*.

Once the flexible form of boat or twist boat has been reached, a series of low energy conformational changes may occur. The energy barrier between the twist boats 3 and 5, through the boat 4, is only about 7 kJ mol^{-1} corresponding to an extremely rapid interconversion at most accessible temperatures. These twist boat–boat interconversions take place *without bond angle changes*. The only changes are in torsion angles and these interconversions are called *pseudorotations*.

In any ring flip where axial and equatorial positions are exchanged it is *necessary* to go through *either* a boat *or* a twist boat (but not both) since the transition state ($\mathbf{C}_s$ or $\mathbf{C}_2$) are associated with only one chair form. (Models!) However, as the pseudorotation barriers are so small it is inevitable that molecules will remain in the high energy 'well' for a short time and undergo a series of 3, 4, 5 interconversions before returning to one chair or another. As 1 and 6 are indistinguishable there is an equal probability that either will be formed from 3, 4 or 5.

Except at very low temperatures the chair–chair interconversion is very rapid, and axial and equatorial hydrogen atoms cannot be distinguished. At any given time less than about 0.1% of cyclohexane molecules are in a flexible form.

7.2.2 Monosubstituted cyclohexanes

There is only one distinguishable chair form for cyclohexane, but let us consider the case for a monosubstituted cyclohexane where one hydrogen atom has been substituted by a ligand, R or X. Now there will be two distinct cyclohexanes, one with the substituent axial and the other with the substituent equatorial. It can readily be established from models that these two forms may be interchanged by ring flips. The axial and equatorially substituted forms of any monosubstituted cyclohexane will have different internal energies as the substituent can lie in one of two different positions with respect to the rest of the molecule. If the two conformers have different energies there will be an equilibrium between the two chair forms.

The consequence of equatorial substitution is illustrated in Figure 7.9. In this case there are no significant extra steric interactions. The equatorial X-group is antiperiplanar to the appropriate methylene groups which is the preferred conformation.

Figure 7.9 Equatorially substituted cyclohexane.

Axial substitution presents a different picture (Figure 7.10, p. 88). The axially substituted X-group has two extra gauche interactions with the neighbouring methylene groups, and there are also extra *1,3 interactions* with hydrogen atoms (Figure 7.10(c)). This simple picture leads to the prediction that equatorial substitution should predominate, as is indeed the case.

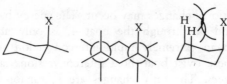

Figure 7.10   Axially substituted cyclohexane.

For most substituents the equilibrium between axial and equatorial substituted chair forms is quite rapid at normal temperature. Variable temperature nmr spectroscopy is the method of choice for studying the equilibrium.
At low temperatures (0° to −120° depending on the substituent) both forms
can be observed separately in the mixture and the equilibrium constant, and
hence $\triangle G^{\ominus}$, calculated. The equilibrium constant K, is defined as,

$$K = \frac{[\text{axial conformer}]}{[\text{equatorial conformer}]}$$

An interesting experiment was carried out by Jensen in 1969. At −150° he
was able to separate the diastereoisomeric chlorocyclohexanes by fractional
crystallization. The axial and equatorial forms were separately redissolved in a
suitable solvent, and their nmr spectra recorded at −150°. Each isomer gave a
single spectrum, uncontaminated by its diastereoisomer. On warming the
solutions to room temperature the equilibrium mixture of axial and equatorial
chlorocyclohexanes was produced from each single isomer. This is an
example of conformational isomerism at 25° and configurational isomerism at
−150°. This experiment is also a practical demonstration of the 'thought
experiment' on p. 53 of Chapter 5.

The equilibrium values of $\triangle G^{\ominus}$, as determined from such nmr experiments, have been put to practical use by Winstein and Eliel to estimate the
steric bulk of substituents. They argued that the equatorial preference of a
substituent should be directly related to its size because the only significant
factor affecting the equilibrium ought to be the non-bonded interactions
shown in Figure 7.10. Although the steric bulk of a substituent is related to
the van der Waals radii there is not always a direct relationship between the
two. The values of $\triangle G^{\ominus}$, sometimes called A values, were postulated to be a
better measure. Some $\triangle G^{\ominus}$ values for the axial–equatorial conformers of
monosubstituted cyclohexanes are given in Table 7.2, and the relationship
between the free-energy difference and the percentage of equatorial
conformer is shown in Figure 7.11. There are several points to note from the
table. As expected, more highly substituted alkyl groups have a greater effect
on the equatorial preference. The culmination of this effect is the t-butyl
group which has such a marked equatorial preference that the conformation is
effectively 'locked' into that of the equatorial isomer. This has remarkably
useful consequences for stereochemical analysis, as shown in the next section.

The $\triangle G^{\ominus}$ values for the OH and NH$_2$ groups are of interest because the groups appear to be different sizes in different solvents. The explanation of this effect is that solvation in hydrogen-bonding solvents does effectively make the ligands bigger, thereby increasing the equatorial preference.

Another kind of monosubstitution is the replacement of a ring CH_2 group by a heteroatomic grouping, such as O, NH or S. Replacement by NH or O has very little effect on the conformational barrier in cyclohexanes as the molecular geometry and the C-heteroatom torsion barriers are relatively unchanged. The replacement of CH_2 by P, S, Se or Si has the effect of slightly lowering the barrier as the ring becomes more distorted owing to longer bonds and consequently lower torsion barriers.

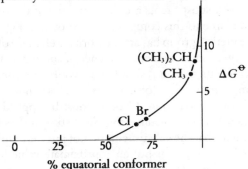

Figure 7.11   The variation of equatorial preference with $\Delta G^{\ominus}$ for monosubstituted cyclohexanes.

Table 7.2. Gibbs free energy differences for axially and equatorially substituted cyclohexanes

Substitutent	$\Delta G^{\ominus}$/kJ mol^{-1}	solvent
CH_3	6.7–7.5	—
C_2H_5	6.7–9.2	—
$HC(CH_3)_2$	7.5–10	—
$C(CH_3)_3$	24	—
C_6H_5	11	$(C_2H_5)_2O$
$CO_2C_2H_5$	5–5.8	C_2H_5OH
$OOCCH_3$	1.5	87% C_2H_5OH/H_2O
COO^-	9.7	H_2O
OH	3.3	75% CH_3COOH/H_2O
	3.8	H_2O
	1.7	CS_2
NH_2	5.0	Aprotic solvents
	6.7	Protic solvents
Br	2.9	—
Cl	2.1	—

89

7.2.3 Disubstituted cyclohexanes

The stereochemical possibilities on disubstitution (both groups different) increases from two chair conformers on monosubstitution to no less than 21 different disubstituted chair forms, (which includes 1,1, 1,2, 1,3 and 1,4 substitution, conformers, enantiomers and diastereoisomers). We shall not catalogue the possibilities; with models, and patience, they may all be discovered; we shall concentrate on interesting conformational effects.

The Winstein–Eliel A values have been used semiquantitatively to predict the conformational preferences of conformational isomers. The process can be illustrated by using cis-4-methylcyclohexanol as an example. The prefix cis indicates that both substituents are the same side of the ring plane. The two conformers available to this compound are shown in Figure 7.12. Table 7.2 shows that the methyl group has an equatorial preference of about 7 kJ mol^{-1}, and the hydroxyl group a similar preference of about 2 to 3 kJ mol^{-1}. If the A values are additive then 7.12(a) should be favoured by approximately 4–5 kJ mol^{-1}. The measured value for $\triangle G^{\ominus}$ is -5.0 kJ mol^{-1} which is in good agreement with the prediction. Caution must be applied to the use of this additivity principle since it assumes that disubstitution has very little effect on ring geometry *and* that the substituents do not interfere with each other. Both are questionable assumptions but an approximate value for the equilibrium constant can often be obtained in this way.

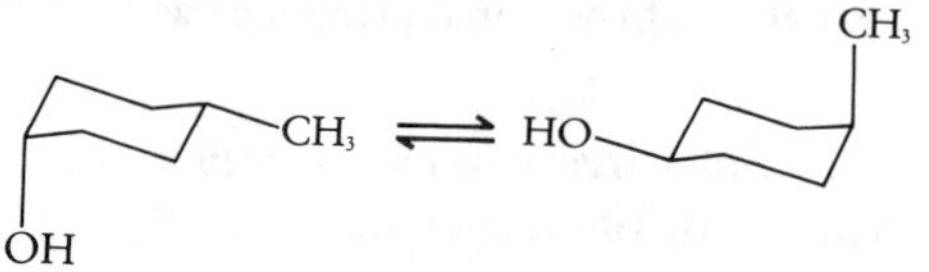

Figure 7.12   Conformers of cis-4-methylcyclohexanol.

The fact that t-butylcyclohexane exists almost exclusively in the equatorial form can be used to advantage. The stereochemical effects of ring substitution on reactivity may be investigated using t-butyl substituted derivatives. The two diastereoisomers of 4-t-butylcyclohexyl ethanoate (acetate) can be isolated pure and free from each other (Figure 7.13). The hydrolysis of the *trans*-isomer is found to be much more rapid than hydrolysis of the cis-isomer under identical conditions. This is one example of the general principle that attack at equatorial positions is easier than attack at axial positions. The axial substituents are much more hindered sterically and approach of a reactant is therefore also hindered. Barton developed this principle by studying steroids and in doing so laid the foundation for the modern stereochemical correlations between structure and reactivity.

If the t-butyl substituent is capable of locking the ring conformation so that the t-butyl group is always equatorial what is the conformation of *trans*-1,3-di-t-butylcyclohexane? Any *trans*-1,3-conformation in the chair form has one group equatorial and the other group *axial*. It proved possible to suggest an answer to this problem by measuring the thermodynamic parameters for

90

the equilibrium between the diastereoisomeric *cis* and *trans*-1,3-di-*t*-butylcyclohexanes. This can be achieved at high temperatures in the presence of a palladium catalyst. The equilibrium and stereochemical consequences are shown in Figure 7.14. As expected, the *trans*-isomer has the higher enthalpy (by about 25 kJ mol^{-1}) but the interesting result is that the *entropy* of the *trans* isomer is also higher by about 20 JK^{-1} mol^{-1}. Flexible cyclohexane conformations have higher entropies than the rigid chair. Examination of models suggest that the conformation of *trans*-1,3-di-*t*-butylcyclohexane is a twist boat as shown in Figure 7.14(c). In that conformation the two *t*-butyl groups are so positioned that Pitzer strain is minimized.

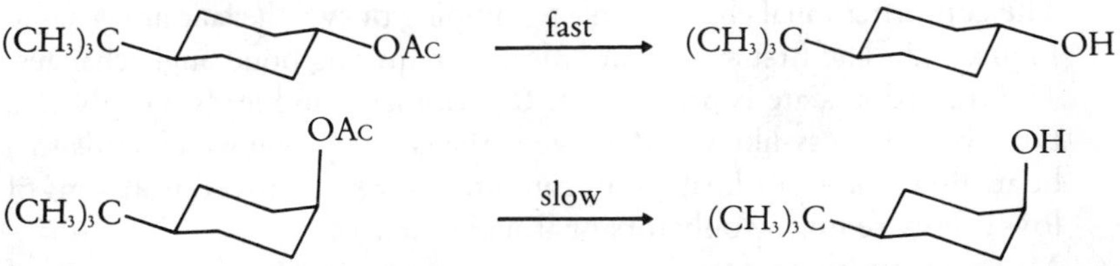

Figure 7.13   Relative rates of hydrolysis of *cis* and *trans*-4-*t*-butylcyclohexyl acetate.

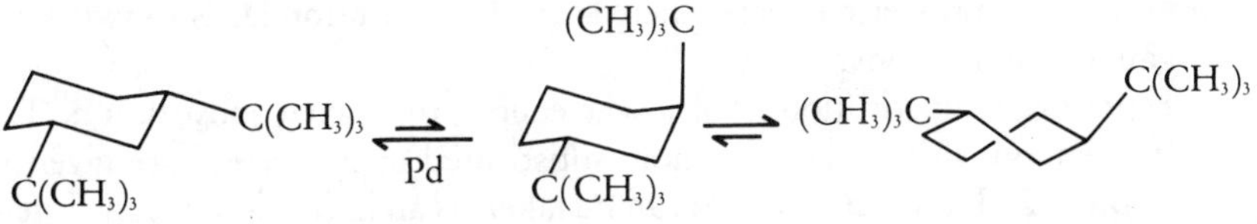

Figure 7.14   Conformations of 1,3-di-*t*-butylcyclohexane. The *trans*-isomer is postulated to have a twist boat conformation.

A confirmation of this idea comes from a study of the all-*cis* isomer of 1,4-di-*t*-butylcyclohexane-2,5-diol in which *intramolecular* hydrogen bonding is observed in the infrared spectrum. This intramolecular interaction between the hydroxyl groups can *only* be obtained in a boat or twist boat conformation. Figure 7.15 illustrates the effect. Model-making, once again, is a useful method of demonstrating stereochemical interactions. A twist boat conformation can be achieved in which the hydroxyl groups are within hydrogen-bonding distance *and* the two *t*-butyl groups are pseudoequatorial (sometimes called *teq* for twist equatorial).

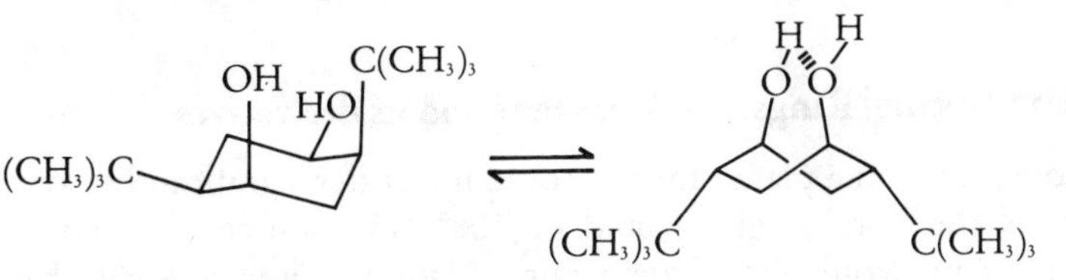

Figure 7.15   Intramolecular hydrogen bonding in *cis*-1,4-di-*t*-butylcyclohexane-2,5-diol.

7.2.4 Summary of Section 7.2

1. The lowest energy conformation of cyclohexane is a rigid chair in which all bonds are staggered. In the chair conformation there are two types of hydrogen atoms called axial and equatorial.

2. Axial and equatorial positions can be interchanged by a chair–chair ring flip.

3. Flexible conformations of cyclohexane are also possible. A *boat* conformation has two completely eclipsed C—C bonds and is at a local energy maximum.

4. A twist boat conformation has no eclipsed bonds and is at a local energy minimum. It exists in two enantiomeric forms.

5. The conformational changes on ring flipping of cyclohexane are given in Figure 7.7. The first step is an *inversion* requiring bond angle changes. The transition state is probably of C_2 symmetry and leads directly to a twist boat. A less-likely alternative is the C_s, envelope which leads to a boat. Boat and twist boat forms can interconvert by pseudorotations of low energy requiring only torsional angle changes.

6. Monosubstitution of cyclohexane leads to an equilibrium mixture of equatorial and axially substituted conformers. Equatorial substitution gives no extra steric effects whereas axial substitution leads to two extra gauche interactions.

7. Equatorial substitution is of lower energy than axial substitution. The free energy differences for monosubstituted cyclohexanes are given in Table 7.2. These $\triangle G^{\ominus}$ values can be taken as an indication of substituent size (Winstein–Eliel).

8. *t*-Butyl substituents effectively lock the cyclohexane ring so that they are equatorial.

9. The $\triangle G^{\ominus}$ values for monosubstitution can be used to get a rough estimate of the conformational preference of disubstitution.

10. Cyclohexane rings locked into one chair form by an equatorial *t*-butyl group can be used to study reactivity. Axial substituents are more hindered than equatorial and similarly approach by reagents is more hindered at axial positions.

11. When two *t*-butyl groups are placed on a cyclohexane ring so that one of them would be axial in a chair conformation, strain is relieved by the adoption of flexible conformation, probably a twist boat.

7.3 The Normal Rings: Cyclopentane and its derivatives

Cyclopentane, in a planar form, should not suffer significantly from Baeyer strain as the bond angles would be 108°. The source of strain in planar cyclopentane would be Pitzer strain. A model clearly shows that planar cyclopentane is fully eclipsed with adjacent bonds in the unfavourable synperiplanar relationship. Cyclopentane adopts a puckered conformation to relieve Pitzer strain, although this is necessarily accompanied by an increase in Baeyer strain. The conformation of cyclopentane is that in which strain is optimized between Baeyer and Pitzer contributions.

There are two possible, high-symmetry, low-energy conformations for cyclopentane, both of which have analogues in the cyclohexane series. The C_s symmetry envelope and C_2 half-chair (twist boat) conformations of cyclopentane are given in Figure 7.16. The hydrogen atoms are labelled according to their relationship with the average ring plane. The symbols a and e again represent axial and equatorial positions and i refers to *isoclinal* positions in which each hydrogen atom is more-or-less equally disposed above and below the ring plane. Isoclinal hydrogen atoms are not all indistinguishable in the envelope, as models will readily confirm, but are equivalent in the half-chair form.

Both the envelope and half-chair are flexible forms that are readily inter-convertible by pseudorotations. There are *no* energy maxima or minima on the cyclopentane profile; only symmetry maxima and minima. Each flexible form of cyclopentane is of constant strain and pseudorotations take place by a wave-like motion around the ring by a series of out-of-plane deformations. It is not necessary to pass through the planar ring to achieve half-chair/envelope interconversions. There are no energy maxima or minima because each pseudorotation causes extra Pitzer strain at one bond that is exactly counterbalanced by a decrease at another bond. There are ten energetically indistinguishable half-chair conformations, and the same number of in-distinguishable envelope conformations for cyclopentane. Conformational analysis is therefore difficult for cyclopentanes.

The degeneracy can be lifted by suitable substitution. Replacement of a methylene group by a heteroatomic group, such as O or NH, induces a distinct preference for the half-chair conformation in which the heteroatom is at the unique position (that bearing the isoclinal hydrogen atoms in Figure 7.16). In this conformation two pairs of H—H eclipsings have been removed and the half-chair is 13 kJ mol⁻¹ more stable than the envelope in the case of tetrahydrofuran $(CH_2)_4O$.

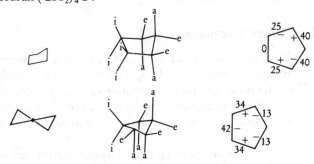

Figure 7.16 The envelope *(top)* and half-chair *(bottom)* conformations of cyclopentane.

In the previous section we described the definite conformational preferences of disubstituted cyclohexanes. Disubstituted cyclopentanes also show a conformational preference but this is rarely very significantly marked. For example *cis*-1,3-dimethylcyclopentane is only 2 kJ mol⁻¹ more stable than the *trans* isomer which compares with about 8 kJ mol⁻¹ for the appropriate cyclohexanes. It is probable that the most stable conformation of the 1,3-dimethylcyclopentanes is the envelope, illustrated in Figure 7.17, in which both methyl groups are equatorial.

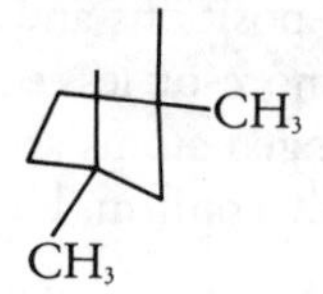

Figure 7.17 The more stable envelope conformation of 1,3-dimethylcyclopentane.

7.4 The Normal Rings: Cycloheptane

Cycloheptane, C_7H_{14}, in common with the other rings studied, is puckered. As the number of atoms in a ring increases so does the complexity of its conformational behaviour. For cycloheptane there are two families of conformations called again, boat and chair. For each family there are two high symmetry forms; a classical and a twist form. The chair family is of lower energy than the boat by about 30 kJ mol⁻¹. One family is converted to the other by an inversion, and conformational changes within each family are by pseudorotations. The four high symmetry forms of cycloheptane are given in Figure 7.18.

Figure 7.18   Conformations of cycloheptane.

7.5 Medium Rings: Cyclooctane

Puckered cyclooctane can take up no less than seven high symmetry forms that can interconvert through inversions, pseudorotations or combinations of the two. There are no strain-free conformations available to cyclooctane, and both Baeyer and Pitzer strains are quite large in all conformations. The most stable form of cyclooctane is probably the boat–chair conformation shown in Figure 7.19.

One of the more interesting aspects of cyclooctane conformation is that this is the smallest ring in which *transannular* interactions may occur. The conformation shown in Figure 7.20 is the boat–boat conformation in which the carbon skeleton corresponds to a fragment of the diamond lattice. A model of this conformation will show that two hydrogen atoms are competing for the same region of space. This conformation therefore has a fairly high energy. Such transannular interactions can be quite severe in cyclooctane and the medium rings.

Figure 7.19   The boat–chair conformation of cyclooctane.

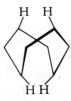

Figure 7.20   The transannular interactions in boat–boat cyclooctane.

7.6 Small rings

Cyclopropanes can have no conformational interest. Any 3 membered ring is necessarily planar and fully eclipsed and therefore has a rigid and unique structure.

Cyclobutanes can, and do, adopt non-planar conformations. The relief of Pitzer strain on ring deformation is accompanied by an increase in Baeyer strain as the bond angles close from 90° in planar cyclobutane to about 86° in the bent form. Representations of bent cyclobutane are given in Figure 7.21. Owing to the high angle strain in cyclobutanes, models are best made with flexible straws which allow conformational deformations to be made easily.

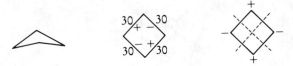

Figure 7.21   Butterfly cyclobutane.

There are two degenerate 'butterfly' conformations for cyclobutane that interconvert through an inversion. The cyclobutane ring is the only alicyclic ring that changes conformation through a one-step inversion. All other rings exchange conformations through a series of steps, often involving both inversions and pseudorotations. The barrier to ring inversion in cyclobutane is 6.2 kJ mol^{-1} and the transition state is the planar four membered ring.

Substitution in the ring has a predictable effect on the inversion barrier. Substituents that favour larger bond angles tend to flatten the ring and therefore lower the inversion barrier. Some examples are given in Figure 7.22.

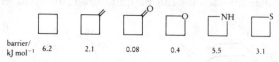

| barrier/ kJ mol^{-1} | 6.2 | 2.1 | 0.08 | 0.4 | 5.5 | 3.1 |

Figure 7.22   Inversion barriers in some cyclobutane derivatives.

7.7 Polycyclic compounds

7.7.1 Introduction

Polycyclic compounds are those that contain more than one ring in the same molecule and are very common among the naturally-occurring compounds. There are four main classes of polycyclic compounds. One has two, or more rings in which each ring is independent and connected to the other(s) by a single bond or bridging atoms. As each ring can act independently no new stereochemical features are introduced. A second class is the *spiro* compounds in which two rings are joined at a single atom. These compounds are of configurational interest and are mentioned in other chapters. The third class is that containing two rings fused together at adjacent atoms. These are called *fused rings*. We shall examine two examples of these. Finally, two or more rings joined at non-adjacent atoms are called *bridged* rings and we examine one example, norbornane.

The interest in fused and bridged rings is two-fold and arises from the effects on conformation and attendant changes in reactivity.

7.7.2 Norbornanes: bridged rings

Norbornane (bicyclo[2,2,1]heptane) is shown in Figure 7.23. The norbornane skeleton appears frequently in naturally-occurring compounds, and particularly in monoterpenes.

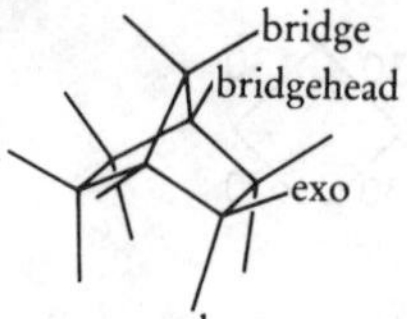

Figure 7.23   Norbornane.

The bridging atom in norbornane effectively locks the six membered ring into a boat conformation. Although norbornane has strict C_{2v} symmetry (complete eclipsing in the 1,3 and 5,6 bonds) substitution does allow a certain conformational freedom so that the boat is distorted towards a twist conformation. Models show how a small twist can be attained.

One particularly interesting aspect of norbornane is the effect of its

96

structure on reactivity. Substitution at C-2 can lead to four distinct compounds; an *exo*-substituted compound, an *endo*-substituted compound and an enantiomer of each. The *exo* and *endo*-substituted forms are diastereoisomers as they cannot be interconverted by conformational changes. In general compounds with *exo*-substituents are lower in energy than their *endo* diastereoisomers, as there is less steric hindrance in the former. This is reflected in the preferential approach of reactants to the *exo* side of the molecule in the norbornane derivatives norcamphor and norbornene,

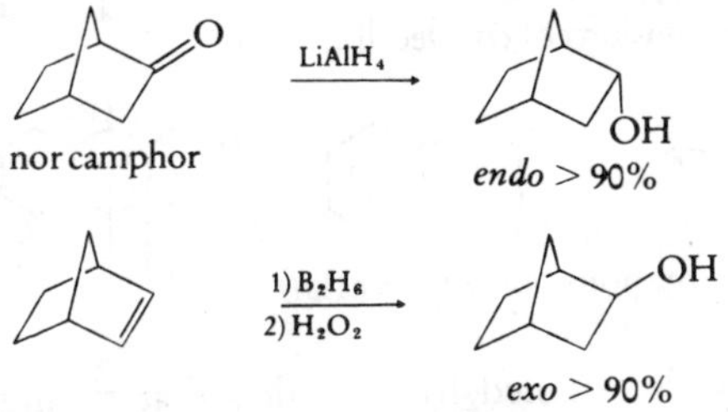

nor camphor

LiAlH₄

endo > 90%

1) B₂H₆
2) H₂O₂

exo > 90%

In these examples the reagents both attack from the *exo*-side and diastereoisomeric products are obtained from the different starting materials.

Substitution at the *bridge* can so hinder *exo* approach that *endo* attack becomes preferred. The reduction of camphor is an example of preferential *endo* attack.

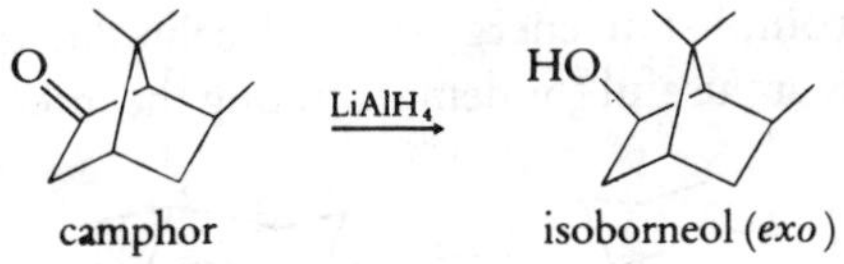

camphor

LiAlH₄

isoborneol (*exo*)

7.7.3 Fused-rings: the decalins
One of the predictions of the Sasche theory of puckered cyclohexanes was that there should be two diastereoisomeric decalins (bicyclo[4,4,0]decanes). In the 1920's Hückel isolated the two forms, called *cis* and *trans* decalin, as shown in Figure 7.24.

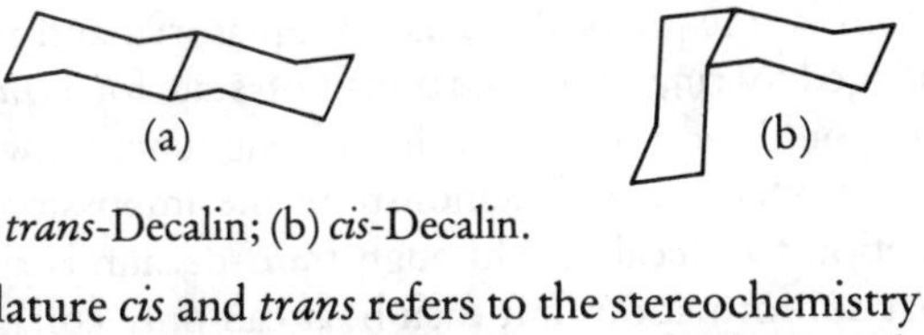

(a)   (b)

Figure 7.24   (a) *trans*-Decalin; (b) *cis*-Decalin.

The nomenclature *cis* and *trans* refers to the stereochemistry of ring fusion and is equivalent to the nomenclature for 1,2 disubstituted cyclohexanes. *cis*-Decalins are fused at an equatorial and an axial position on each ring, and it should be noted that the equatorial position with respect to one ring is axial with respect to the other, as illustrated in Figure 7.25. *trans*-Decalin has a ring fusion that is equatorial to both rings as shown in Figure 7.26.

An examination of molecular models shows that *cis*-decalin has three more gauche, butane-like interactions than *trans*-decalin. The *cis* isomer is found to be 11 kJ mol⁻¹ less stable than the *trans*-isomer, principally resulting from the extra gauche interactions.

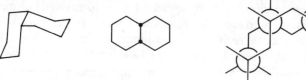

Figure 7.25   Representations of *cis*-Decalin.

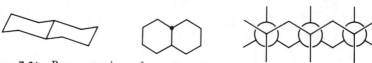

Figure 7.26   Representations of *trans*-Decalin.

It is also readily determined from models that *cis* and *trans*-decalin are diastereoisomers because they cannot be interconverted by bond rotations.

The conformational behaviour of each isomer will now be investigated. *cis*-Decalin exists as an equilibrium between two enantiomeric conformations (Figure 7.27) that are readily interconverted by ring flips. The activation free energy, $\triangle G^{\ddagger}$ for this process is 54 kJ mol⁻¹ compared with 43 kJ mol⁻¹ for cyclohexane. This relatively low barrier for two ring flips is probably a result of the increase in ground state energy of *cis*-decalin due to the extra gauche interactions. Models are useful for demonstrating the interconversion.

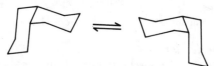

Figure 7.27   The two, enantiomeric conformation of *cis*-decalin.

trans-Decalin, on the other hand, is called a *rigid* structure. A substituent, or hydrogen atom is unequivocally axial or equatorial and its environment cannot be exchanged by ring flips. The transition state for a *trans*-decalin ring interconversion would require a *trans*-diaxial ring junction which cannot be attained. Again, model use will demonstrate the impossibility of a *trans*-diaxial ring junction for decalin. Although *trans*-decalin is rigid, each chair can attain a boat conformation, but each boat can only collapse back to the original chair. The populations of the boat–chair and boat–boat conformations are very small indeed.

The rigid conformation of *trans*-decalin has been used in the same manner as *t*-butyl groups to study the relative rates of equatorial and axial substitution. The hindered approach to axial positions is again demonstrated by the ester hydrolyses shown in Figure 7.28.

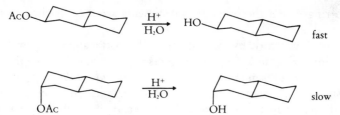

Figure 7.28   The approach to axial positions in *trans*-decalin is more hindered than that to equatorial positions.

Heteroatom substitution can, in certain cases, dramatically decrease the barrier to interconversion of *cis* and *trans*-decalin. If a nitrogen atom occupies a bridgehead position as in Figure 7.29, *cis* and *trans* forms cannot be isolated. The rapid interconversion of the *cis* and *trans* forms in this example is a result of the easy inversion of configuration at nitrogen which removes the need for a *trans*-diaxial transition state.

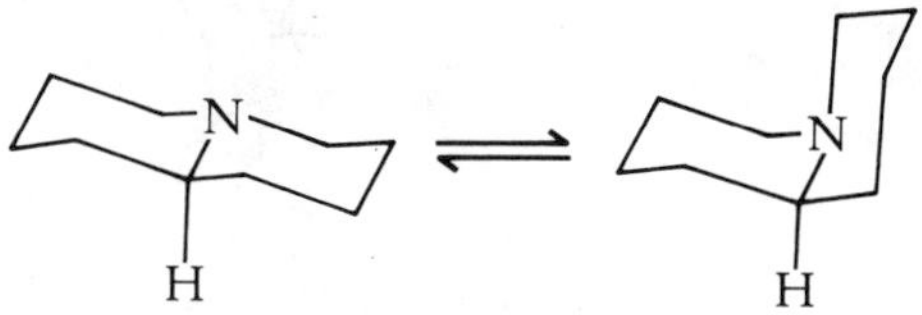

Figure 7.29   The interconversion by ring flipping of azadecalin.

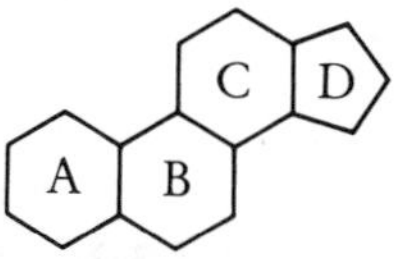

Figure 7.30   The steroid nucleus.

7.7.4 Fused-rings: steroids

The steroids are ubiquitous, naturally-occurring compounds with the four ring skeleton shown in Figure 7.30. There are four cyclohexane rings and one cyclopentane ring. Steroids can belong to one of two families that resemble the decalins at the A and B rings. The *cholestane* family (7.31(a)) has a *trans*-decalin like AB ring system and the *coprostane* series has a *cis*-decalin AB rings (Figure 7.31(b)). In both series the BC and CD ring junctions are all *trans*.

The presence of a series of rings confer a rigidity on *both* series. Although boat conformations can be attained, even the coprostane series cannot exchange axial and equatorial positions by ring flips.

Steroids are chiral and, as found in nature, usually occur as only one of many possible isomers. The representations in Figure 7.31 gave absolute configurations. It is customary, when dealing with steroids, to label one face

99

of the ring α and the other β. Substituents are then labelled as say, 3-β-equatorial, etc.

The original work by Barton on conformational analysis was with the cholestane series, and attack by reagents was shown to be from the relatively unhindered α-side, and this was designated the *rule of the rear*.

Positional preference of substituents parallels that in cyclohexanes. Steroids are frequently substituted at C-3 and, as would be predicted from cyclohexanes a 3-β-equatorial substituent is more favoured than a 3-α-axial substituent. When X or Y in Figure 7.31(a) is OH, and there is a chemical equilibrium between the 3α and 3β forms the equilibrium is,

$$3\beta(OH) \rightleftharpoons 3\alpha(OH)$$
$$95\% \qquad 5\%$$

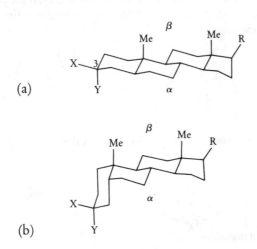

(a)

(b)

Figure 7.31   (a) The cholestane series of steroids; (b) the coprostane series of steroids.

7.8 Summary of Sections 7.3–7.7

1. Cyclopentane exists as an equilibrium between a set of C_2 half-chair conformations and a set of C_s envelope conformations. The interconversion, which has no energy barrier, is by pseudorotations.
2. Cycloheptane can belong to one of two families of puckered structures. The chair family has both chair and half-chair forms and is lower in energy by 30 kJ mol^{-1} than the boat family which has classical and twist forms also. Interconversions within families are by pseudorotations and from one family to another by inversions.

100

3. Cyclooctane has a complex conformational set, of which the lowest energy form is probably the boat–chair. Transannular interactions can be important in certain conformations.

4. Cyclopropane is rigid and has no interest from a *conformational* viewpoint.

5. Cyclobutane exists in two degenerate butterfly forms that interconvert by inversion with $\triangle G^{\ddagger}$ 6.2 kJ mol^{-1}.

6. Norbornane has a bridged ring. The six membered ring is fixed in a boat conformation. *Exo*-attack is favoured over *endo*-attack, unless the bridge position is substituted when *endo*-attack is preferred.

7. There are two diastereoisomeric decalins. *cis*-Decalin has two enantiomeric forms interconvertible by ring flips, $\triangle G^{\ddagger}$ 54 kJ mol^{-1}. The *cis*-decalin is less stable than the *trans* isomer by 11 kJ mol^{-1} as there are three more gauche interactions in its most stable conformation.

8. *trans*-Decalin is rigid and cannot undergo ring flips, as a *trans*-diaxial transition state would be required.

9. A bridgehead nitrogen atom allows ready interconversion of the *cis* and *trans* forms through nitrogen inversion.

10. Steroids have a four ring system and can have either a *cis* AB junction (coprostanes) or a *trans* AB junction (cholestane).

11. Steroids from both series are rigid.

12. Attack at steroids by chemical reagents is from the unhindered α-side (the rule of the rear).

Problems and Exercises

1. Use models to help you to discover the number of different chair conformations of 1,3-dichlorocyclohexanes, which conformations, if any, are chiral and which are interconvertible by ring flips? What conclusions can be drawn about the number of 1,3-dichlorocyclohexanes likely to exist at room temperature?

2. Intra molecular interaction between a carboxylic acid group and an alcohol group gives a lactone

   Use models to help determine which isomer of 1,4-C_6H_{10}(OH)COOH is likely to form a lactone? Draw the structure of the final lactone product.

3. Identify the different hydrogen environments in a cyclohexane twist boat and suggest names for them based on cyclopentane twist conformations and chair cyclohexanes.

4. How many different conformations can you draw for ?

CHAPTER 8

Chiral compounds

8.1 Introduction

A molecule without reflection symmetry is chiral, and it necessarily follows that chiral molecules are not superimposable on their mirror images. This is not simply an academic point but one of fundamental importance in chemistry and biology. Naturally-occurring organic compounds, such as sugars and proteins are predominantly chiral and their chirality is essential to their function. In this chapter we shall examine some chiral molecules and some of their properties, following the introduction to chirality in Chapters 3 and 4.

The elements that characterize chiral molecules are *chiral centres, chiral axes, chiral planes* or a combination of these elements. We shall examine first molecules that contain *one* of these elements and then extend the analysis to molecules with two or more chiral centres. Finally the properties and discrimination of chiral molecules will be studied.

8.2 Molecules with one chiral centre

This discussion is limited, for reasons of space, to molecules in which the chiral centre is a carbon atom, although organic molecules with chiral heteroatoms, such as N, P, Si and S, are common.

A molecule Cabcd is *necessarily* chiral if $a \neq b \neq c \neq d$ and none of the ligands a–d are themselves chiral. (In certain cases a chiral ligand can confer mirror symmetry on a molecule Cabcd as explained in Section 8.5.2). There are no other restrictions, the ligands can be alkyl, aryl, heteroatomic or a mixture of these. All molecules with a single chiral atom belong to point group C_1.

The α-amino acids, $RCH(NH_2)COOH$ are an important class of chiral compound and many contain only one chiral centre. These are the building blocks for peptides and proteins. With the exception of glycine,

$H_2C(NH_2)COOH$, all twenty or so protein amino-acids have the same *relative configuration* as shown in Figure 8.1. The *absolute configuration* of chiral compounds is often specified by the use of *Fischer projections* (Figure 8.1) in which the disposition of ligands about the central carbon atom is indicated by their position on the cross. Ligands on horizontal branches are defined as being above the paper plane whereas ligands on vertical branches are below the paper plane. Note that the rotation of a Fischer projection through 90° either clockwise or anticlockwise produces the *enantiomer* and rotation through 180° leaves the configuration *unchanged*. This should be verified by comparison of models with Fischer projections.

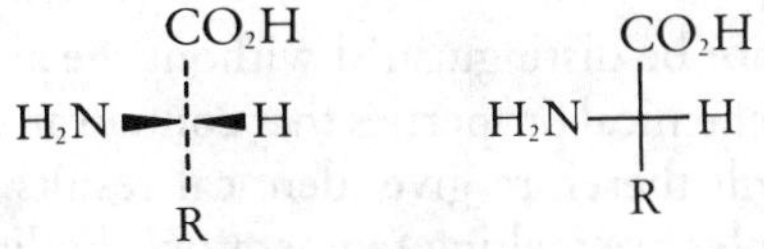

Figure 8.1 The Fischer projection (*right*) of the naturally occurring α-amino acids with the absolute configuration shown (*left*).

The central carbon atom in chiral compounds is often called an *assymmetric,* or chiral carbon atom (no symmetry elements) although it is, of course, the whole molecule that is chiral, and gives rise to chiral properties.

In addition to Fischer projections the recommended *R,S* nomenclature (Appendix 1) can be used to specify the absolute configuration of chiral compounds. An older system devised by Emil Fischer in the late nineteenth century is still frequently used to describe amino-acids and their derivatives. That is the *D,L* system that relates the configuration to that of a standard compound *S*-glyceraldehyde, shown in Figure 8.2, which was given the label *L*-glyceraldehyde by Fischer. As there was no method available for the determination of configuration at that time Fischer arbitrarily gave *L*-glyceraldehyde the configuration shown in Figure 8.2. In the 1950's it was established that Fischer's choice was, in fact, correct.

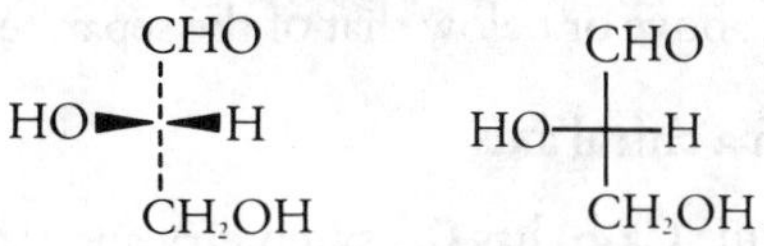

Figure 8.2 Representations of *L*-(*S*)-glyceraldehyde.

There is no necessary connection between the D, L and R, S conventions. The *D, L* convention is based on the *chemical* transformation of compounds leading to either *D* or *L* glyceraldehyde. Compounds that can be correlated with *L*-glyceraldehyde belong to the *L*-series and analogously for the *D*-series.

Protein amino acids belong to the *L*-series which gives the local symmetry for the NH_2, COOH and H ligands shown in Figure 8.1. Members of the L-series can have either R or S configurations depending on the priority of the R ligand. The use of the D, L system continues because it demonstrates the essential similarity between all of the protein amino acids whereas the R, S system does not always do so. It is, nevertheless, inevitable that the D, L-system will fall into disuse because modern synthetic transformations are becoming so sophisticated that there are already examples where amino acids can be related to either D or L glyceraldehyde depending on the route chosen.

The R, S system is therefore recommended for use, although a knowledge of the D, L system is essential for understanding much of the chemical and biological literature.

Enantiomers cannot be distinguished without the aid of an external chiral agent. Physical and chemical properties that contain an element of symmetry, *scalar* properties, will therefore give identical results for each of a pair of enantiomers. Examples are melting temperature, boiling temperature, solubility in achiral solvents, bromination by Br_2, reduction by $LiAlH_4$ etc.

A chemical reaction that produces a chiral compound Cabcd will always give exactly equal amounts of each enantiomer if there is an element of symmetry in the reaction, for example,

$$PhCH_2CH_3 \xrightarrow{\text{N-bromosuccinimide}} PhCHBrCH_3$$
$$50\%R, \ 50\%S.$$

An equimolar mixture of enantiomers is called a *racemic modification*. In the vapour and liquid phases a racemic modification behaves as an ideal mixture of enantiomers. In the crystalline state there are two different possibilities for the form of a racemic modification. Each enantiomer may crystallize independently to give a mixture containing discrete crystals of each enantiomer which is called a *racemic conglomerate*. As this is an *eutectic mixture* the melting temperature of a racemic conglomerate is always lower than that of the separate enantiomers. In other cases both enantiomers co-crystallize to give a different crystal structure in which each enantiomer is found in the unit cell. Such crystals are called *racemates* and can have melting temperatures either above or below that of the separate enantiomers.

8.3 Molecules with a chiral axis

A tetrahedral molecule Ca_2b_2 has $\mathbf{C}_{2v}$ symmetry and is achiral (Figure 8.3(a)). If the tetrahedron is elongated along one of the symmetry axes of the archetypal tetrahedron the symmetry can be lowered. Figure 8.3(b) shows the effect of elongating along one C_2 axis so that $\mathbf{C}_2$ symmetry results. The elongation axis is called the *chiral axis* and it is the asymmetrical or disymmetrical disposition of ligands *about this axis* that produces chirality.

The first $\mathbf{C}_2$ compound with a chiral axis that was resolved into enantiomers was the allene shown in Figure 8.4. The $\mathbf{C}_2$ symmetry is the maximum allowed for a chiral allene, $abC = C = C\,cd$. (Compounds in which $a = d \neq c = b$ have

$\mathbf{C_2}$ symmetry and are chiral. Compounds with $a \neq b \neq c \neq d$ are also chiral with $\mathbf{C_1}$ symmetry (no symmetry elements), and an example is given in Figure 8.5. Numerous chiral allenes are now known and have been resolved.

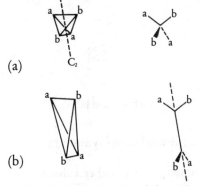

(a)

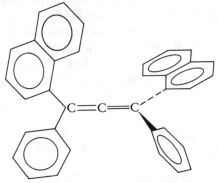

(b)

Figure 8.3 Production of a chiral axis by notional elongation of a tetrahedron.

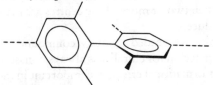

Figure 8.4 The first allene to be resolved into enantiomers.

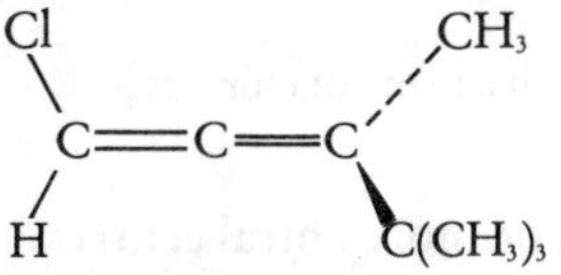

Figure 8.5 An asymmetric allene.

Other compounds with a chiral axis are the biphenyls, Figure 8.6, exhibiting atropisomerism (Chapter 6).

Figure 8.6 Biphenyl compounds have a chiral axis.

105

The helicenes mentioned in Chapter 4 are a special case of axial chirality. These are sometimes classified separately having *helical symmetry* (Figure 8.7).

Figure 8.7   Helical symmetry (a left handed helix).

8.4 Molecules with a chiral plane of symmetry

Certain chiral compounds possess neither a chiral centre nor a chiral axis but contain a *chiral plane*. *trans*-Cyclooctene (Figure 8.8(a)) is such a compound. The molecule is chiral because there are two asymmetric arrangements of the tetramethylene bridge, $(CH_2)_4$, either above or below the alkene *plane*, defined by the double bond and its four directly attached atoms.

Other examples with chiral planes are generally associated with bridged aromatic rings such as the paracyclophanes shown in Figure 8.8(b). The substituted benzene ring and its directly bonded atoms specify the chiral plane.

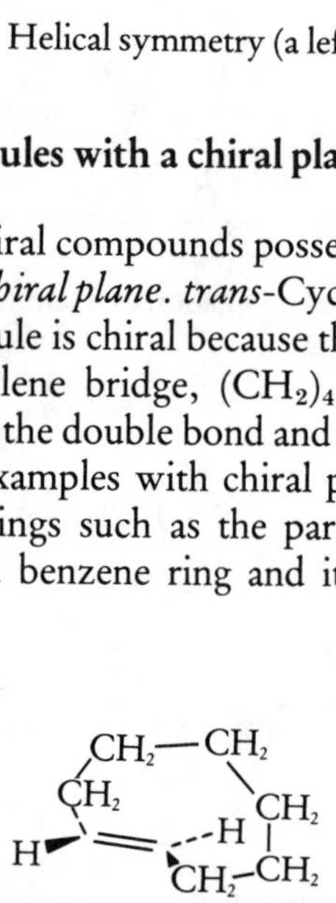

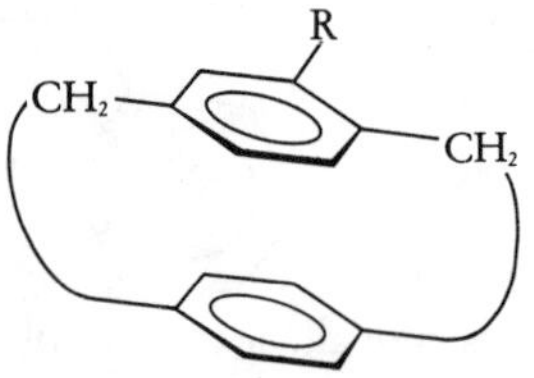

Figure 8.8   Compounds with a plane of chirality.

8.5 Molecules with two or more chiral centres

8.5.1 Introduction

As the number of chiral centres (or other elements) in a molecule increases, the number of stereoisomers increases rapidly. In an acyclic compound with n *different* chiral centres there are 2^n *stereoisomers* that are arranged as 2^{n-1} *pairs of enantiomers*. When two or more of the centres are the same the number of stereoisomers is reduced.

No new principles are needed to study compounds with more than one chiral centre. The criterion for chirality is always non-superimposability on a mirror image. Conformational effects are important in these compounds and the guidelines drawn up in Chapter 5 apply here.

8.5.2 Molecules with two chiral centres

A molecule with two different chiral centres has four stereoisomeric forms, consisting of two diastereoisomeric pairs of enantiomers.

Isoleucine is an α-amino acid with two different chiral centres. All four stereoisomers are shown in Figure 8.9. Fischer projections have again been used to specify configuration in a conventional way. Fischer projections carry *no conformational information* and molecules are always given an eclipsed form, whatever their preferred conformation. The labelling of the molecules is straightforward: 1 is labelled $2R, 3R$ by applying the sequence rules to each chiral atom separately. The molecules 1 and 2 with opposite configurations at both carbon atoms are enantiomers, as are 3 and 4. The relationship between all other pairs of molecules is diastereoisomeric. A new term can be introduced here; diastereoisomers that differ only in the configuration at *one* carbon atom are called *epimers*. All of the diasteroisomeric pairs for 1–4 are therefore also epimers. The advantage of this term is only really apparent when studying reactions where only one of many chiral centres may be affected, for example C-3 in the steroid series.

The number of stereoisomers of a molecule is reduced when one or more pairs of stereoisomers has reflection symmetry. The classical example of such a molecule with two chiral centres is tartaric acid, studied by Louis Pasteur. There are only three forms for tartaric acid as shown in Figure 8.10. The molecules 5 (R, R) and 6 (S, S) are enantiomers and are diastereoisomeric with the R, S molecule 7. Compound 7, formerly called racemic acid, is not chiral, owing to the presence of a mirror plane of symmetry. Molecules containing chiral centres, that are not themselves chiral, are called *meso* compounds. Inspection of models will verify that 7 is achiral.

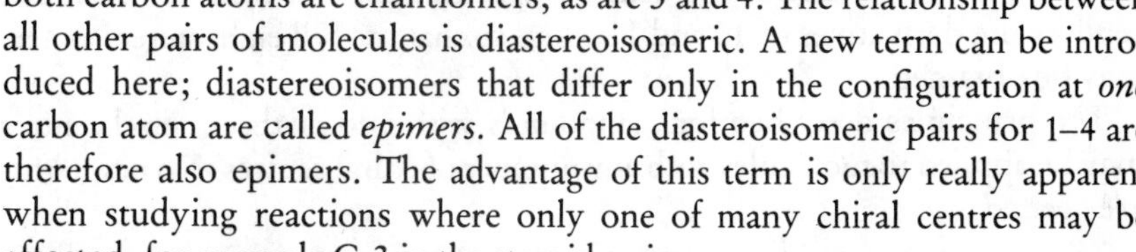

^{1}COOH	COOH	COOH	COOH	COOH
H—^{2}C◄NH$_2$	H——NH$_2$	H$_2$N——H	H——NH$_2$	H$_2$N——H
H—^{3}C◄CH$_3$ $\equiv$	H——CH$_3$	H$_3$C——H	H$_3$C——H	H——CH$_3$
C$_2$H$_5$	C$_2$H$_5$	C$_2$H$_5$	C$_2$H$_5$	C$_2$H$_5$
2R, 3R	2R, 3R	2S, 3S	2R, 3S	2S, 3R
1	1	2	3	4

Figure 8.9 Fischer projections of the stereoisomers of isoleucine.

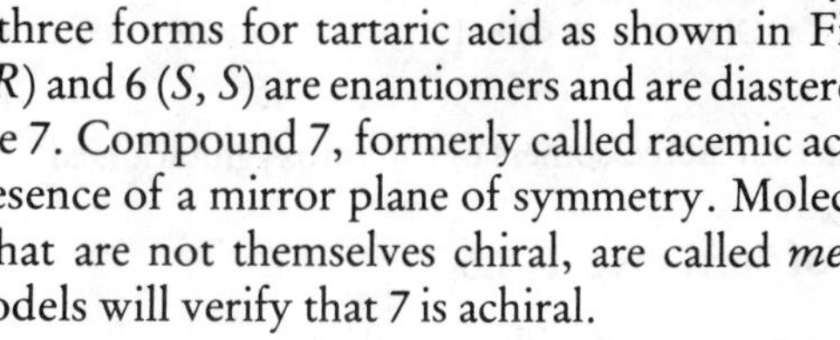

COOH	COOH	COOH
H——OH	HO——H	H——OH
HO——H	H——OH	H——OH σ
COOH	COOH	COOH
R, R	S, S	R, S
5	6	7

Figure 8.10 The three forms of tartaric acid.

8.5.3 Molecules with more than two chiral centres

Compounds with three different chiral centres have eight stereoisomeric forms, consisting of four diastereoisomeric pairs of enantiomers.

Just as compounds with two enantiomeric chiral centres are *meso*, so other compounds with two or more pairs of symmetrically disposed enantiomeric centres are also *meso*. Trihydroxyglutaric acid gives an example as shown in Figure 8.11. Molecules 8 and 9 are enantiomers and are diastereoisomeric with the two diastereoisomeric *meso* compounds 10 and 11. Models are ideal for demonstrating the mirror plane of symmetry in 10 and 11 and can also be used to confirm that they are diastereoisomers. It can also be verified that substitution at *one* of the carboxylic acid groups, to make an ester for example, renders the derivatives of 10 and 11 chiral so that the full eight stereoisomers may be obtained.

The principles described for compounds with two or three chiral centres may be applied to molecules with any number of chiral centres. No new terms are needed.

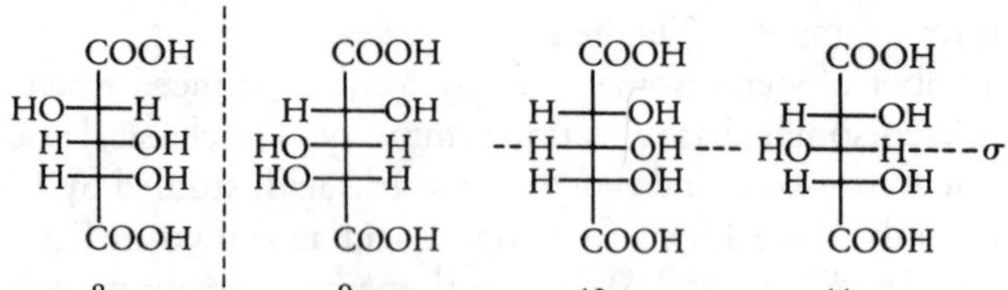

Figure 8.11   The four stereoisomers of trihydroxyglutaric acid.

8.6 Summary of Sections 8.1–8.5

1. A chiral molecule is one without reflection symmetry and three elements can characterize chiral compounds; chiral centres, chiral axes, or chiral planes. A helix is a special case of axial symmetry, although some authors classify helices separately.
2. The absolute configuration of enantiomers may be denoted by the use of Fischer projections. Ligands on horizontal branches are, by convention, above the paper plane, and those on vertical branches are below the paper plane. For long chain molecules with > 1 chiral centres the Fischer projection is always drawn in the all-eclipsed conformation.
3. In formulae, absolute configuration is denoted by the R, S or D, L systems. The D, L system relates configurations to that of R or S glyceraldehyde by chemical transformations. The R, S convention is unambiguous and should always be used.
4. Chemical reactions in the absence of a chiral influence always give equal amounts of enantiomers. Equimolar quantities of enantiomers are called racemic modifications. Crystals of racemic modifications can be racemic conglomerations of separate crystals or racemates which incorporate both enantiomers into the unit cell.

5. A compound with *n* different chiral centres has 2^n stereoisomers with 2^{n-1} pairs of enantiomers. Compounds with two or more identical chiral centres have fewer stereoisomers.
6. Epimers are diastereoisomers differing in the configuration at one carbon centre only.
7. *meso* compounds contain two or more chiral centres but are themselves achiral, through the possession of a mirror plane.

8.7 The manifestations of chiralty: The characterization of enantiomers by chiroptical methods

Enantiomers can only be discriminated in the presence of a chiral agent. *Plane polarized light* is a chiral agent. If normal light is passed through a polarizing filter, such as polaroid, the *electric field vector*, associated with the propagation of light, oscillates in a *single plane* perpendicular to the direction of propagation (Figure 8.12).

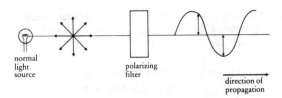

Figure 8.12   Schematic representation of the plane polarization of light.

This plane polarized light can be described as being composed of two, enantiomeric, helical waves circulating around the axis of propagation in opposite directions. (Examples of axial, or helical chirality). Each helix has an associated electric field vector, and the two waves are in phase so that at any time the contributions of the two electric field vectors to the propagation cancel *except in the plane of propagation*. This combination is shown schematically in Figure 8.13.

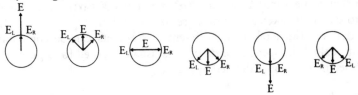

Figure 8.13   The combination of two helical electric field vectors to give plane polarized light. Propagation is perpendicular to the page.

When plane polarized light is passed through a solution containing a chiral compound there is a diastereoisomeric interaction. The chiral molecules will

109

refract one helix more than the other; each component will have a different *refractive index* in a chiral medium. If the chiral molecules are all of the same configuration, or one enantiomer is in excess, the plane of polarization will be *rotated* relative to the original plane. This process, called *circular birefringence* is illustrated in Figure 8.14. The direction and extent of rotation may be measured by using a second polarizing filter as the detector. This phenomenon forms the basis for the analysis of enantiomers by their *optical rotations*. An apparatus called a *polarimeter* is used to study optical rotations by passing *monochromatic* plane-polarized light (usually the sodium **D** line at 589 nm) through a cell of fixed path length and measuring the rotation. A rotation in a clockwise direction, relative to an observer, looking through the solution towards the light source, is denoted as positive (+) and an anticlockwise rotation is negative (−).

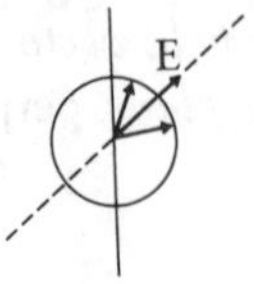

Figure 8.14 Circular birefringence. The original plane of polarization is given by the vertical line. One component of the light is refracted more than the other, so that the resultant plane is rotated relative to the original plane.

An enantiomer with a positive rotation is sometimes described as *d* (dextro) (*not* to be confused with *D* !) and one with a negative rotation as *ℓ* (laevo).

If a solution of a pure enantiomer, in an achiral solvent has a rotation of $+\alpha°$ then its enantiomer, under identical conditions has a rotation of $-\alpha°$. Enantiomers that rotate the plane of polarized light are said to be *optically active*. Optical activity is a sure sign of chirality but the absence of rotation does not necessarily mean the absence of chirality. Some chiral molecules have a vanishingly small rotation and are therefore chiral but optically inactive.

The extent of rotation can vary linearly or non-linearly with concentration. Non-linear variation is a sign of molecular association (such as hydrogen-bonding) affecting the structures of molecules in solution. The variation is linear, except in the case of molecular association. Care must therefore be taken that a measured rotation of $\alpha°$ is not $n360° \pm \alpha$; measurements are therefore usually taken at two or more concentrations.

So that rotations taken on different instruments can be comparable, a quantity called the *specific rotation* [α] is usually quoted for a pure enantiomer. Specific rotation is defined by,

$$[\alpha]_\lambda = \frac{\alpha}{c.l}$$

where λ is the wavelength of light used, α is the observed rotation, c is the concentration in $g cm^{-3}$, and l is the cell path length in *dm*. The solvent chosen may affect the conformational equilibrium and hence the rotation, so the

110

solvent is usually quoted too. It is also wise to quote the temperature, as density, conformation and molecular association are all temperature dependent.

As a comparison of the power of rotation the *molar rotation* [Φ] is preferable to the specific rotation (but less frequently quoted), and

$$[\Phi] = [\alpha]_\lambda \, \frac{\mathbf{M}_r}{100}$$

Where $\mathbf{M}_r$ is the relative molar mass quoted in grams.

The optical rotation of mixtures of enantiomers is frequently measured, particularly as a measure of the success of creating a particular stereoisomer. (See Chapter 14.) In such cases the chemical yield and the *optical purity* or *yield* are usually given.

If $[\alpha]_{max}$ is the specific rotation of one enantiomer and $[\alpha]$ the measured specific rotation the optical purity P is defined by,

$$P = \frac{[\alpha]_{max}}{[\alpha]}$$

In general,

$$P = \frac{[d - \ell]}{[d + \ell]}$$

where the ratio $[d - \ell]/[d + \ell]$ is called the *enantiomeric purity*, and $[d]$ and $[\ell]$ are the concentrations of the $d(+)$ and $\ell(-)$ enantiomers, respectively. The quantities P and $[d - \ell]/[d + \ell]$ are zero for racemates and either -1 or $+1$ for pure enantiomers, depending on the enantiomer taken as standard. Mixtures can have values anywhere between -1 and $+1$. Note that the sign of rotation at one wavelength does not carry any intrinsic information about the absolute configuration of an enantiomer. The rotation of one enantiomer may be positive at one wavelength and negative at another. The other enantiomer mirrors this behaviour exactly so that at *any* wavelength, under the same conditions, enantiomers have equal and opposite rotations.

A plot of the variation of [α] with [λ] is called an *Optical Rotatory Dispersion* curve (ORD or ord). The refractive index for right and left circularly polarized light varies strongly with wavelength, particularly in the region of a *chromophore* (a molecular feature that has an electronic transition in the observed region e.g. C=O, phenyl, conjugated alkenes). Two types of chromophore are observed. An inherently *disymmetric chromophore* is one that is chiral and lies on an axis or in a plane of chirality. An *asymmetrically or disymmetrically perturbed chromophore* is itself achiral but is influenced by neighbouring chirality.

The variation of [α] with [λ] can belong to one of two types. In the *absence* of any chromophore a *simple ORD curve* is observed. Figure 8.15 (p. 112) shows two simple ORD curves. An enantiomer that gives a steady increase in rotation, (not necessarily involving a change of sign) from high to low wavelengths is said to exhibit a *positive Cotton effect* and the opposite is a *negative Cotton effect*.

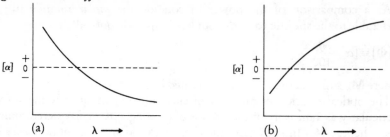

Figure 8.15   Simple ORD curves (a) positive Cotton effect; (b) negative Cotton effect.

When a chromophore of either type is present, an *anomalous Cotton effect* is observed, as illustrated in Figure 8.16. A positive Cotton effect is, in this case, defined by a curve in which, from high to low wavelength, a *peak* is met before a trough. The opposite is again called a negative Cotton effect.

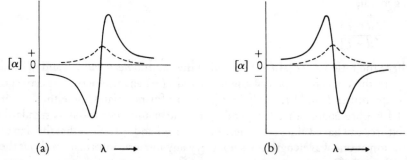

Figure 8.16   Anamalous ORD curves (a) positive Cotton effect; (b) negative Cotton effect. The dotted curve represents the absorption spectrum of the chromophore.

The *amplitude* of the Cotton effect is the difference in specific rotation between the highest maximum and the lowest minimum. It is a useful guide to the presence of inherently dissymmetric chromophores as they may cause amplitudes of even hundreds of thousands of degrees, whereas remote achiral chromophores may only produce amplitudes of a few degrees. The amplitude of the Cotton effect is a measure of the power of the optical rotation and can be correlated with molecular dissymmetry.

In the next section the Cotton effect is applied to the deduction of the configuration of chiral molecules.

The Cotton effect is also manifested in a more simple manner in a technique called *Circular Dichroism* (CD or cd). Circular Dichroism depends on the fact that chiral molecules with chromophores have slightly different *absorptions* for right and left circularly polarized light. The consequence of this

112

differential absorption is shown in Figure 8.17. If, for example, the left circularly polarized light is absorbed more strongly than the right component then the resulting electric field vector no longer traces out a simple wave pattern. Instead, a highly flattened, elliptical helix is produced. In general the major axis of this ellipse is much greater than the minor axis, and for the purposes of measurement may be treated as plane polarized light. When this differential absorption or *molar ellipticity* is plotted against wavelength, a curve closely related to the electronic absorption spectrum is observed. This curve can be either positive or negative as shown in Figure 8.18. This is again called the Cotton effect but has the advantage of simplicity of interpretation compared with the ORD curve which may produce a broad and complicated curve if more than one chromophore is present. CD bands are narrow and usually non-overlapping even in the presence of more than one chromophore. Again, the amplitude of the CD Cotton effect is related to molecular dissymmetry.

As with the other *chiroptical* quantities, such as specific rotation and ORD curves, enantiomers have mirrored CD curves.

Figure 8.17 *Left* plane polarized lights; *right* elliptically polarized light produced from plane polarized light by differential absorption.

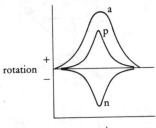

Figure 8.18 Circular dichroism curves *p* is a positive Cotton effect; *n* is a negative Cotton effect and *a* is the absorption spectrum.

8.8 The determination of configuration

8.8.1 Chiroptical methods

There is currently only one completely reliable method for determining *absolute* configurations directly and that is anomalous X-ray diffraction *(vide infra)*. All other methods depend, to a greater or lesser extent, on the correlation between a compound of known configuration with one of unknown configuration. This correlation can be through chemical or physical properties.

The simplest and most trivial use of the chiroptical methods for the determination of configuration is simply to compare the sign and magnitude of the specific rotation of the compound of unknown configuration with the specific rotation of a sample of the same compound with a known configuration.

The more sophisticated use of chiroptical techniques applies to compounds of unknown configuration. The sign and amplitude of the CD or ORD Cotton effect can be used comparatively particularly for closely related series of compounds. The most successful use of the Cotton effect is empirically-based and called the *octant rule*. This rule was first developed for carbonyl compounds and has been extended to aromatic compounds. The precursor of the octant rule was the observation that, in chiral cyclohexanones the introduction of equatorial halogen substituents had no effect on the Cotton effect whereas axial substituents could change its sign. It was subsequently found that the contribution to the sign of the Cotton effect from groups or atoms in other parts of the molecule could also be predicted.

The octant rule for ketones operates as follows. A coordinate system is set up so that the $C=O$ bond defines one axis, the x axis in the usual convention; the plane containing the $C=O$ atoms and the two atoms bonded to the carbonyl carbon constitutes the xy plane and the yz plane passes through the CO bond. An example is given in Figure 8.19.

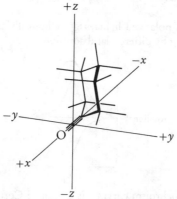

Figure 8.19 The orientation of cyclohexanone in the cartesian coordinate system that defines the octants used in the octant rules.

The effect of any substituent on the sign of the Cotton effect can be obtained by multiplying together the *signs* of its coordinates. This is most readily understood through examples. A 2-axial substituent is covered first; such a substituent has the cartesian coordinates $-x, +y, -z$, which gives a positive contribution on multiplying the signs together. Therefore a 2-axial substituent gives a positive contribution to the Cotton effect. The substituent effect can alternatively be described in terms of *octants*. A 2-axial substituent observed along the CO bond towards the rest of the molecule is designated as being in the right-lower-rear octant. Each octant is associated with a sign as determined above. A table can be made of the contribution of a substituent in any octant.

Similarly a 6-axial substituent in the $-x$, $-y$, $-z$ or left-lower-rear octant has a negative effect. Both 2- and 6-equatorial substituents lie almost in the xy plane and therefore have zero as at least one coordinate. Any group lying in any plane makes a negligible contribution to the sign of the Cotton effect.

To a *first* approximation the effect of different substituents is supposed to be equal and additive. Therefore for simple molecules the octant rule can lead to the absolute configuration of a compound by comparison with analogous ketones of known configuration. In some cases where the configuration is known the octant rule can give information about the conformation, although this is *more* difficult.

Where there are several substituents having opposite contributions to the Cotton effect, more care in interpretation is needed. A series of empirical, semi-quantitative contributions to the Cotton effect for different substituents has been devised and these may help in the deduction of configuration. Axial halogens generally have a dominant effect compared with most other substituents.

In examples where the sign (and/or amplitude) of the Cotton effect varies significantly from the calculated values a change of conformation, perhaps from a chair to a boat as is often suggested. For this and other reasons the Cotton effect must be used with caution, particularly as some substituents, notably fluorine are unpredictable in their effect.

8.8.2 Configuration by chemical correlation

The D, L system of configurational nomenclature has been described for amino acids. This is a relative nomenclature and was based on Emil Fischer's arbitrary (but correct) assignments of absolute configuration of the enantiomeric glyceraldehydes.

The basis of this, and other chemical correlations is the chemical transformation of a compound of unknown configuration into a compound of known configuration. This correlation must take place through a series of stereochemically unambiguous steps. The most unambiguous correlations are those that involve transformations at atoms other than the chiral centre. In later chapters we describe reaction stereochemistry but it is necessary here to describe the possible stereochemical consequences of a substitution reaction on a chiral molecule CabcX,

$$XCabc \rightarrow YCabc$$

The two stereochemically precise routes shown in Figure 8.20 (p. 116) are *retention* of configuration, in which the local symmetry of Cabc is preserved, and *inversion* of configuration where the local symmetry of the Cabc group in YCabc is opposite to that in the starting material. A reaction may take place by either route or a combination of the two. A reaction giving 50% inversion and 50% retention is called a racemization reaction.

If the stereochemical outcome of a particular transformation is precise *and* well-established, such reactions can be used in the correlation of configuration. Many such precise reactions are now known but should only be used for configurational correlation when there is no doubt as to their stereochemistry.

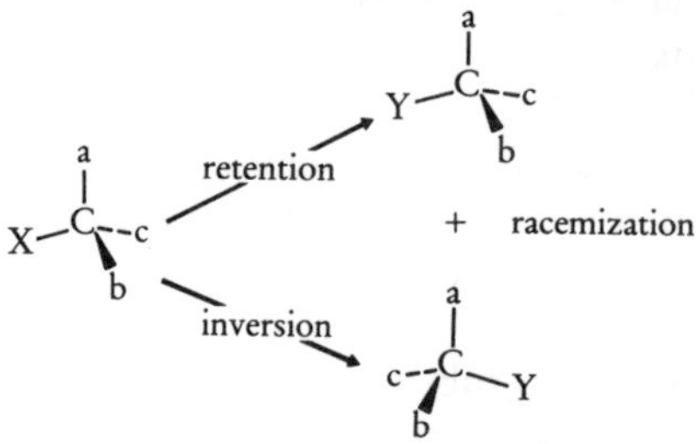

Figure 8.20   Stereochemical possibilities of substitution.

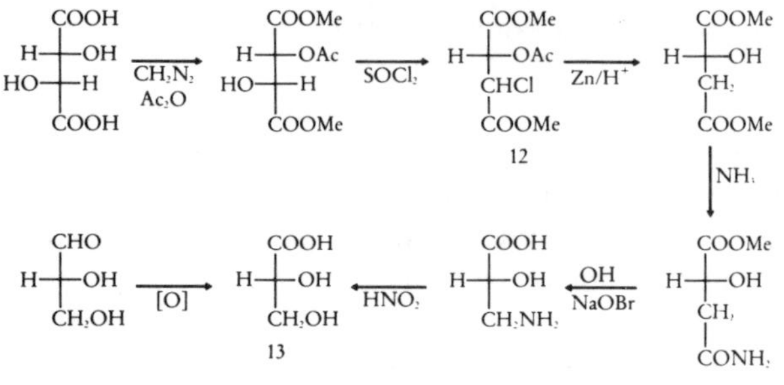

Figure 8.21   The chemical correlation of (+)-tartaric acid with R-glyceraldehyde.

An excellent, if rather involved, example of the correlation of configuration through chemical modification is shown in Figure 8.21, and shows the correlation of (+)-tartaric acid with R-glyceraldehyde. The chemical constitution of the tartaric acid was known and both carbon atoms are necessarily of the same configuration. (The compound with different configurations at each carbon atom is the *meso*-tartaric acid.) Therefore, it was sufficient to establish the configuration of only one chiral carbon centre. The chemical transformations shown in the figure constitute a multistage degradation during which the chirality at one carbon atom was preserved and the chirality at the other carbon was removed. Note that there was only one stage with a transformation of indeterminate stereochemistry and that was the reaction leading to compound 12. The stereochemistry of this reaction is immaterial as the chirality at that carbon atom was destroyed in the next stage. The final correlation with R, (D)-glyceraldehyde required the mild oxidation of the aldehyde group in glyceraldehyde to give the carboxylic acid 13 which was the

116

final product in the tartaric acid degradations. The two samples of compound 13 obtained by the different routes were shown to have specific rotations of the same sign and magnitude.

Chemical correlations for allenes have also been devised but depend on a knowledge of reaction stereochemistry.

8.8.3 Absolute configuration by calculations and anomalous X-ray diffraction

Calculations of the optical rotation of a compound can be carried out by using a quantity called *bond polarizability*, which is a measure of the response of electrons in bonds to electric field gradients. These calculations are complex and at present have limited use.

The method of choice for the determination of absolute configuration is an X-ray method called anomalous X-ray diffraction. Normal X-ray diffraction is a scalar method as right and left cannot be distinguished. The phenomenon of anomalous X-ray scattering usually depends on the presence of a heavy atom that can absorb X-rays as well as diffract them. The pattern obtained from this technique depends on the chirality of the molecule and allows the unequivocal assignment of absolute configurations. The technique is too complex to describe in detail here.

8.9 The separation of enantiomers

Any method for separating enantiomers depends on the presence of a chiral agent giving rise to diastereoisomeric interactions.

One method that uses our ability to distinguish left from right, by explicit or implicit reference to a chiral object, such as a hand, is the Pasteur method of *mechanical separation*. This involves the manual separation of racemic conglomerates. Pasteur separated the enantiomeric R, R and S, S-tartaric acid crystals using tweezers and a microscope! Mechanical separation is rarely possible.

The usual methods for separation of enantiomers are chemical or occasionally biological. The basis of the chemical separation of a racemic mixture is the reaction of a racemic mixture R + S with a chiral agent X to give a pair of diastereoisomers RX and SX. As diastereoisomers have different chemical and physical properties they can be separated by conventional means. The final stage is the regeneration of R and S. This process may be illustrated,

$$R + S \rightarrow RS + SX \overset{\nearrow\ SX \longrightarrow S}{\underset{\searrow\ RX \longrightarrow R}{\text{separation}}}$$

The chiral *resolving agent* X must be enantiomerically pure or nearly so, or separation cannot be complete. If X is itself a racemic mixture of X and $\overline{X}$, four compounds will be formed and it will be impossible to separate RX from its enantiomer S$\overline{X}$ and SX from R$\overline{X}$.

If the racemic mixture is acidic, reaction with a chiral base can give diastereoisomeric salts that are separable by fractional crystallization or distillation. Similarly bases may be resolved with chiral acids.

Biological separation is efficient once the right system has been discovered but is rather drastic as it depends on the preferential consumption of one enantiomer. An example is the degradation of *D*-amino acids by the enzyme *D*-amino oxidase (extracted from cobra venom). *L*-amino acids are unaffected. The following reaction can take place.

D-amino acid

　　　+ 　　　　$\xrightarrow[\text{oxidase}]{\text{D-amino}}$ 　　　$RCOCH_2OH + H_2O_2 + NH_3$

L-amino acid 　　　　　　　　　　　　　　　*L*-amino acid

8.10 Summary of Sections 8.7–8.9

1. Plane polarized light is a chiral agent consisting of two opposing helical electric field vectors.
2. Chiral compounds affect the refraction of right and left circularly polarized light to different extents so that the plane of polarization is rotated. One enantiomer will rotate the plane by $+\alpha°$ and the other under identical conditions will rotate the plane by $-\alpha°$.
3. The specific rotation at a given wavelength is,

$$[\alpha]_\lambda = \frac{\alpha}{cl} \text{ and molar rotation } [\Phi] = \frac{[\alpha]\mathbf{M}_r}{100}$$

4. Enantiomers may be specified as + or − at a given λ but this is unrelated to their absolute configurations.
5. A plot of $[\alpha]$ vs λ is called an Optical Rotatory Dispersion or ORD curve; the sign of which depends on molecular asymmetry. A simple Cotton effect is observed in the absence of a chromophore and is positive if $[\alpha]$ increases towards the lower wavelengths. An anomalous Cotton effect is observed in the presence of an inherently dissymmetric chromophore or an asymmetrically or dissymmetrically perturbed chromophore. Positive Cotton effects are those with a peak met before a trough (high to low λ).
6. The amplitude of the Cotton effect can be correlated with molecular asymmetry, and is greatest for inherently dissymmetric chromophores.
7. Circular Dichroism (CD) measures molar ellipticity, or differential absorbance. In the region of a chromophore Cotton effects are observed.
8. The Cotton effect for CD or ORD is used in the correlation of compounds of known configuration with those of unknown configuration. The general method applies to the octant rule where the position of the substituent in the molecule is used to predict its effect on the sign of the Cotton effect.

9. Chemical correlation involves the transformation of a compound of unknown configuration into one of known configuration through a series of stereochemically well-defined steps.
10. Anomalous X-ray diffraction gives absolute configurations unambiguously.
11. Enantiomers can be separated manually in a few cases but chemical methods predominate. A racemic mixture is first converted to a pair of diastereoisomers that can be separated by physical methods. The pure enantiomers can then be regenerated separately.

Problems and Exercises

1. Classify the following molecules as having either, centres, axes or planes of chirality. Where possible, assign the molecules as R or S.

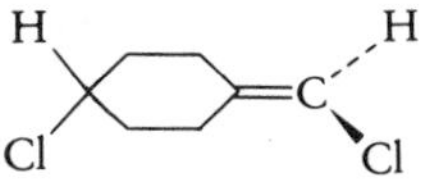

2. The molecule H—$\overset{\displaystyle CO_2H}{\underset{\displaystyle C_6H_5}{\vert}}$—$OH$ can be described as R, $D-$ $(-)$ mandelic acid.
   Distinguish between the various labels and describe how they may be determined and state what information each label contains.
3. The Cotton effect for $2R,SR$-trans-2-chloro-5-methylcyclohexanone is positive in methanol, but negative in octane. The other enantiomer shows equal and opposite behaviour. What phenomena may be responsible for this behaviour? Describe the procedure for estimating the Cotton effect in this compound, there is no need to attempt analysis in detail.
4. How might models be used most effectively in determination of chirality, configuration and estimation of Cotton effects?

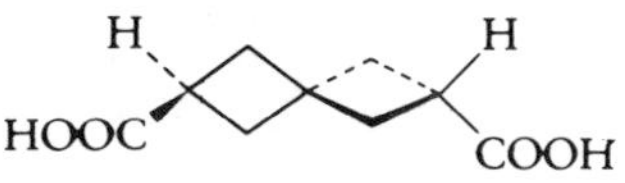

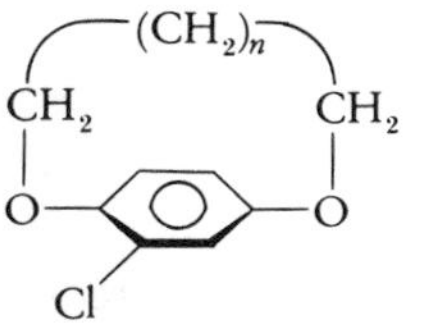

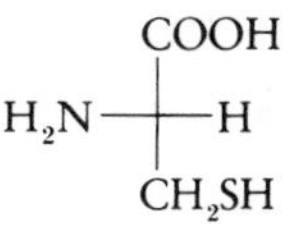

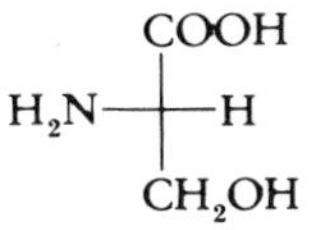

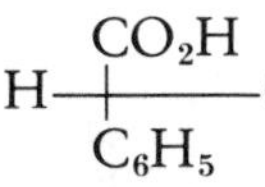

CHAPTER 9

Macromolecular stereochemistry

9.1 Introduction

Macromolecules are 'giant' molecules with molecular weights ranging from around one thousand to hundreds of thousands. There are two types of macromolecules; man-made and natural. Man-made macromolecules are usually described as *polymers* and are made by the process of *polymerization* of *monomers*. The simplest example of a polymer is probably the ubiquitous polyethylene (poly ethene) which is made by bonding together thousands of ethene monomer molecules.

$$n\text{H}_2\text{C}\text{=}\text{CH}_2 \quad \xrightarrow{\text{catalyst}} \quad (\text{CH}_2\text{CH}_2)_n$$

The —CH_2CH_2— unit is called the *repeat unit* and is closely related to the monomer structure.

We shall not be concerned with methods of polymer synthesis but with the structure of the polymers produced. Polymer, and macromolecular structure in general, is a subject of some complexity and experimental difficulties arise in structure determination. A great deal is now known about macromolecular structure and its interest arises from the relationship between form and function. This is a subject where both configuration and conformation are equally important and there is an intimate relationship between the two. This relationship is equally true for the natural macromolecules. We shall briefly discuss two types of naturally occurring macromolecules, polysaccharides derived from glucose, and poly (amino) acids such as peptides and proteins. The polysaccharides, and polyethylene, are derived from a single monomer and are called *homopolymers*. Polyamino acids may be homopolymers, such as polyglycine or *copolymers* made up of a number of different monomers, which is the case for proteins in general.

120

9.2 Hydrocarbon polymers

Most polymers are not simple chemical compounds in the sense of being made up of a number of molecules with identical constitution and molecular weight. Polymers generally consist of a number of long chains of varying length containing the same repeat units. The molecular weight of a polymer is an average for a wide distribution of chain lengths. The physical properties of a polymer are a function of the average molecular weight and for a particular polymer the crystallinity is greatest for the samples with the highest average molecular weight (relative molar mass).

In the crystalline state *polyethylene* chains adopt a regular, linear zig–zag shape as shown in Figure 9.1. The conformation is the lowest enthalpy, all-antiperiplanar conformation analogous to that described in Chapter 6 for hydrocarbons. Regions where large numbers of linear chains are aligned are called *micro-crystalline* regions. Different chains are held together by Van der Waals forces which although individually rather small can become significant over a pair of long chain molecules.

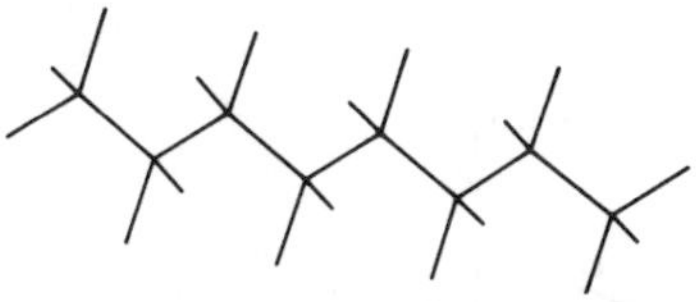

Figure 9.1   The all-antiperiplanar conformation of polyethylene in the solid phase.

In solution the entropy effect becomes important and the chains are more likely to adopt randomized conformations.

The question of configuration is not applicable to polyethylene as there is no possibility of chirality or diastereoisomerism. Configuration and constitution become central to structure when polymers of $RCH{=}CH_2$ are considered. First consider constitution. $RCH{=}CH_2$ could theoretically polymerize in one of two *regioselective* ways (see Chapter 10),

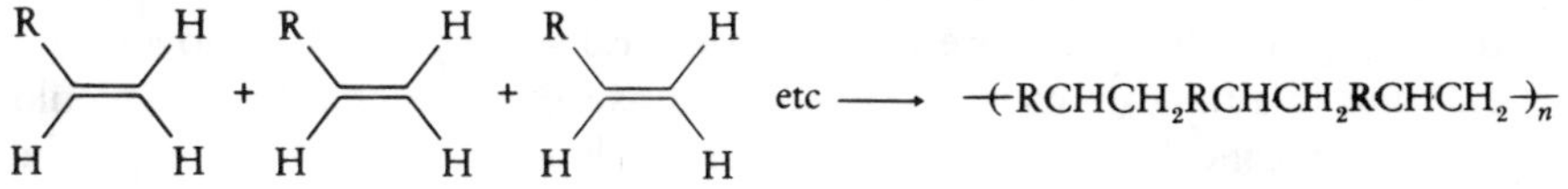

$$\text{etc} \longrightarrow \text{---}(RCHCH_2RCHCH_2RCHCH_2)_n$$

or   (9.1)

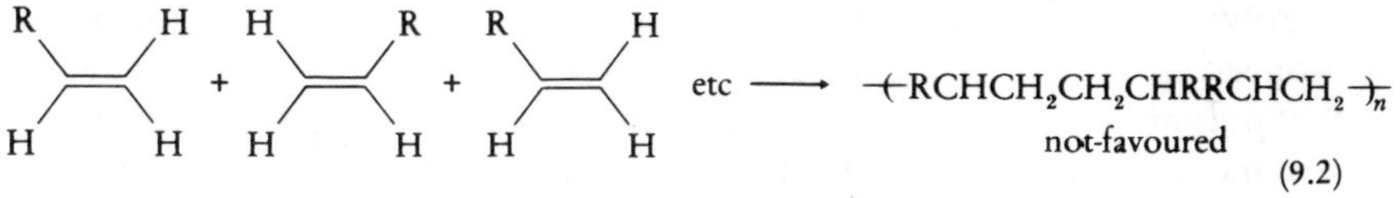

$$\text{etc} \longrightarrow \text{---}(RCHCH_2CH_2CHRRCHCH_2)_n$$
not-favoured
(9.2)

or possibly in a random fashion. In fact the first, regioselective polymerization (equation 9.1) is found almost exclusively for man-made polymers.

121

The question of configuration at the carbon bearing the R group now arises. The three possible stereochemical relationships between the R groups on neighbouring carbon atoms are shown in Figure 9.2. For convenience the chains are shown in the all-antiperiplanar conformations. An *isotactic* polymer X—(RCH$_2$—CH$_2$)$_{\overline{n}}$ Y is one in which all the R groups are on the same side of the molecule. This chain is effectively *achiral*. Ignoring the end groups X and Y, which, even if different, will have very little effect on a very long chain, it can be shown that there is a plane of symmetry running through the centre of the chain at right angles to the direction of the chain.

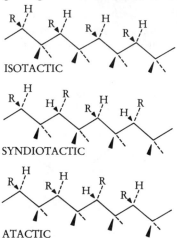

Figure 9.2   Isotactic, syndiotactic and atactic polymers of RCH═CH$_2$.

A *syndiotactic* polymer with R groups alternating regularly between opposite sides of the chain is similarly *de facto* achiral. *Atactic* polymers have a random disposition of R groups on either side of the chain. Each individual atactic chain is likely to be chiral but a sample of an atactic polymer with achiral R groups will not be optically active. There are two reasons for this. As the groups on the chains are so similar, optical activity would be low in any case and in a random chain the contributions from each chiral centre would practically cancel. In addition to this, in a truly random polymer sample the number of chains with a net positive rotation would be counterbalanced by chains with a net negative rotation.

Polymerization technology is now so sophisticated, largely through the pioneering work of E. Ziegler and G. Natta, that conditions can be altered so that polymers of a particular tacticity can be readily produced. In fact atactic polymers are the *least* likely to be formed. Atactic polymers frequently contain large segments of isotactic and/or syndiotactic chains.

It is the tacticity, above all, that determines the conformation of polymer chains. The X-ray structural analysis of syndiotactic polypropylene shows

122

that this polymer, like polyethylene has a staggered conformation shown in Figure 9.3. Inspection of a model of a fragment of such a chain will help to verify that in this all antiperiplanar conformation the methyl groups are not interfering with each other.

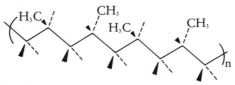

Figure 9.3   Syndiotactic polypropylene in the antiperiplanar conformation.

Isotactic polypropylene is not at its lowest energy in the antiperiplanar conformation as there are interactions between the 1,3-methyl groups similar to those in cyclohexanes. The conformation in which these interactions are minimized is a coiled helix as shown in Figure 9.4. The helix illustrated is a right-handed helix in which the carbon backbone defines the helix. The schematic diagram of Figure 9.4(b) shows the view looking down the helix and illustrates how the methyl groups are pointing away from the coiled backbone, and therefore not interfering with the rest of the molecule. It is advisable to make a model of this helix to get a proper perspective on its three-dimensional structure. A helix is of course chiral but isotactic polypropylene is not optically active because the number of right-handed helices is counterbalanced by the presence of left-handed enantiomeric helices. Rotation about C—C bonds also allows interconversion of right- and left-handed helices. Again, models are ideal for demonstrating the enantiomeric nature of the right- and left-handed helices of isotactic polypropylene. The helices of isotactic polypropylene contain an unusual element of symmetry, a *screw-axis* which is a combination of a translation and a rotation.

Figure 9.4   Schematic representations of the coiled, helical structure of isotactic polypropylene.

The monomer units repeat themselves with a translation every three units. But each monomer is related to its neighbour by a rotation of 120° followed by a translation of one monomer unit either up or down the helix depending on the direction of rotation. This is called a *3-fold* screw axis and can be demonstrated by careful manipulation of models.

An examination of the conformation of the chain in the helix helps with an understanding of its stability. All the bonds are staggered and alternate between antiperiplanar and synclinal arrangements of alternating sign. This minimizes the gauche interactions of the methyl groups so that they are antiperiplanar to one neighbouring bond and synclinal to the other. Calculations confirm that this is the lowest energy conformation.

The atactic form of polypropylene, if it is truly random will adopt an irregular conformation. If it consists of large blocks of iso and syndiotactic regions the sample will consist of small regions of helices and other regions of linear chains.

The structures described for polypropylene are also repeated for other polymers of $RCH=CH_2$, such as polystyrene $-(PhCH-CH_2)_n$. The helices do not necessarily have 3-fold screw axes and helices with orders up to 26 are known.

The chain and helical structures of polymers are usually described as their *secondary* structure, whereas the configuration and constitution define the *primary* structure.

One more example will be sufficient to demonstrate the effect of configuration on hydrocarbon polymer structure. Butadiene, $CH_2=CHCH=CH_2$, can give one of four stereoregular polybutadienes, according to the polymerization conditions.

Polymerization can be in a 1,2 manner,

$$etc \rightarrow -(CH-CH_2-CH-CH_2-CH-CH_2)_n-$$
$$\underset{CH=CH_2}{|} \quad \underset{CH=CH_2}{|} \quad \underset{CH=CH_2}{|}$$

or 1,4

$$etc \rightarrow -(CH_2-CH=CH-CH_2CH_2-CH=CH-CH_2)_n-$$

The 1,2-polybutadienes can be either isotactic or syndiotactic and have helical or linear zig–zag secondary structures in which the $-CH=CH_2$ group replaces the methyl group in polypropylene.

The two 1,4-polybutadienes are differentiated by the configuration about the carbon–carbon double bonds. 1,4-*trans*-Poly-butadiene consists of long linear staggered chains with the conformation and configuration shown in Figure 9.5(a). This is a rubbery substance that is closely related to the natural compound gutta-percha (used for the rubbery insides of golf balls) shown in Figure 9.5(b).

The other 1,4-polybutadiene has a *cis* (Z) configuration at the double bonds and is illustrated in Figure 9.6(a). Natural rubber, shown in Figure 9.6(b) has a very similar structure. Both gutta-percha and natural rubber may be considered to be *terpenes*; derivatives, in a formal sense, of *isoprene* $(CH_3)_2C=CHCH_3$. Both of the *trans*-1,4 polyenes are more crystalline

124

than their *cis* counterparts as the chains are more linear and pack together more easily.

As a final point on hydrocarbon polymers it should be noted that *cross-linking* can effect their mechanical properties immensely. Cross-linking is a process whereby adjacent chains are tied together by chemical bonds. This confers extra rigidity on the structure by preventing adjacent chains slipping over one another and losing 'memory' of their previous position. Natural rubber is synthetically cross-linked by sulphur which forms disulphide bridges between chains at the points where double bonds existed. The process of sulphur cross-linking, called vulcanization, confers such strength on the rather tacky natural rubber that it becomes strong enough for car tyres and many other purposes.

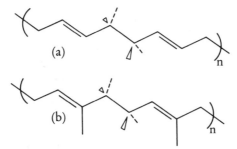

Figure 9.5 (a) 1,4-*trans*-polybutadiene; (b) gutta-percha.

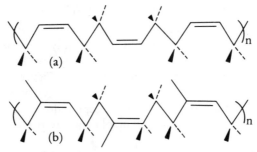

Figure 9.6 (a) 1,4-*cis*-polybutadiene; (b) natural rubber.

9.3 Peptides and proteins

Proteins are made up of thousands of amino acids linked together by amide bonds, formed by condensation of the amino group on one amino acid with the carboxylic acid group of another

$$R-\underset{\underset{H}{|}}{\overset{\overset{NH_2}{|}}{C}}-COOH + H_2N-\underset{\underset{R}{|}}{\overset{\overset{COOH}{|}}{C}}-H \longrightarrow R-\underset{\underset{H}{|}}{\overset{\overset{NH_2}{|}}{C}}-\underset{\underset{O}{\|}}{C}-N-\underset{\underset{COOH}{|}}{\overset{\overset{H}{|}}{C}}-R$$

125

As each amino acid is difunctional the process could, in principle, be extended *ad infinitum*. We shall concentrate on the naturally occurring proteins and peptides that are composed exclusively of *L*-amino acids so the configuration is fixed. There are two other conformational effects that dominate the structure of polyamino acids. The amide unit is always planar and in the preferred *Z* conformation; and given the planar, fixed conformation of the amide bond the only torsional angles that can affect the structure are the C_α—CO angle, ψ, and the $C_\alpha N$ angle Φ. Figure 9.7 illustrates the planar, *Z* nature of the amide group and defines the torsion angles. In the figure both ψ and ϕ have values of $\pm 180°$ by a convention adopted in 1971. A positive rotation is one in which the rear group is rotated in a *clockwise* direction relative to a stationary forward group.

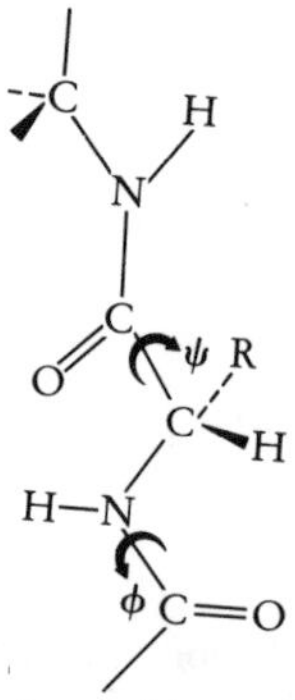

Figure 9.7 The torsional angles ϕ and ψ in polypeptides and proteins.

The effect of the R group in the amino acid can be quite dramatic and various secondary structures are possible for polyamino acids. The simplest structure, possible only for the polymer of glycine (R=H) is the planar, *antiparallel sheet or β structure*, and is shown in Figure 9.8. Long chains of polyglycine are held together by interchain hydrogen bonds. In this example both ϕ and ψ are 180° so that all atoms except hydrogen atoms bonded to the α-carbon are in a single plane, built up of many antiparallel chains.

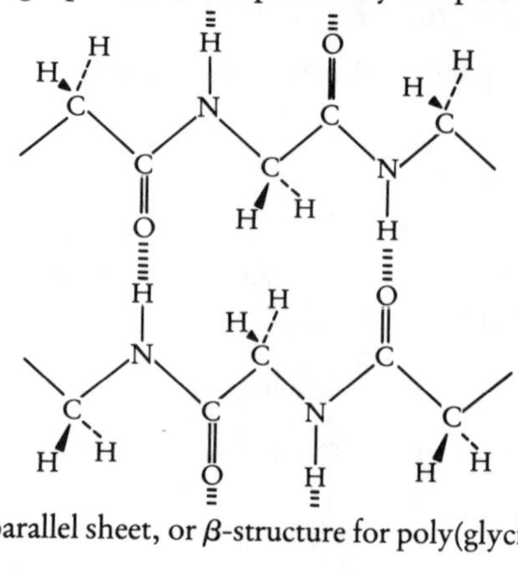

Figure 9.8 The antiparallel sheet, or β-structure for poly(glycine).

When R $\neq$ H there is too much steric crowding of the R groups on adjacent chains to allow the formation of a planar, antiparallel sheet. In this case the more stable sheet structure is the *β-pleated* sheet illustrated in Figure 9.9, in which the values for the torsional angles are about $-140°$ for ϕ and $+145°$ for ψ. The R groups are therefore placed alternately above and below the average plane and steric hindrance is reduced.

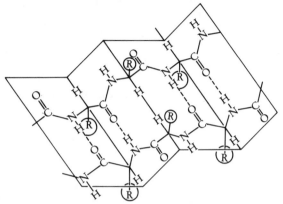

Figure 9.9   The β-pleated sheat structure for poly amino acids.

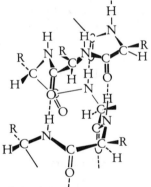

Figure 9.10   A right-handed α-helix structure for poly amino acids.

The pleated sheet structure is one of the two most common, regular structures found in proteins. The other is a helical structure called the *α-helix*. As in the hydrocarbon polymers a poly (amino acid) helix can be right- or left-handed, but as the amino acids of the *L*-series are chiral the two helices are diastereoisomeric and therefore have different internal energies. The more stable of the two is generally the right-handed helix shown in Figure 9.10. In poly (*L*-alanine), R$=$CH$_3$, the torsion angles are $\psi - 47°$ and $\theta - 58°$. The helix is held together by intrachain hydrogen bonds. This helix parallels the hydrocarbon helix in that the R groups are arranged around the outside of the helix to reduce steric repulsion. In an α-helix there are 3.6 amino acid residues for each full turn. Following Figure 9.10 it should be possible to construct a right-handed helix of poly (L-alanine). Note that these are thirteen atoms in each ring formed by hydrogen bonding (use a longer straw for hydrogen bonds than normal bonds), and that the carbonyl groups all point roughly along the helix axis in the same direction towards the oppositely disposed N—H groups. A model shows that in the right-handed α-helix the R group eclipses the NH group. If a left-handed helix is also constructed it can be observed that in that case the R group eclipses the carbonyl group. The extra stability of the right-handed helix is probably associated with the smaller size of the N—H group relative to C=O.

In addition to α-helices, λ-helices with 4.4 amino acid residues are sometimes found in proteins and in very rare cases the ω-helix has been observed with exactly four residues per turn.

9.4 Polysaccharides

Simple sugars have the formula
HC(O)CH(OH)CH(OH)CH(OH)CH(OH)CH$_2$OH.
There are four different chiral centres and therefore 2^4 or sixteen stereoisomers. We are only concerned with polymers of one sugar, glucose. The configuration of the open-chain form of glucose is given in the Fischer projection in Figure 9.11. Sugars are not, however, usually found in the open-chain form but are usually cyclic with an oxygen containing five- or six-membered ring. The five-membered ring series are called *furanoses* and the six-membered ring sugars are called *pyranoses* (as derived from furan and pyran respectively).

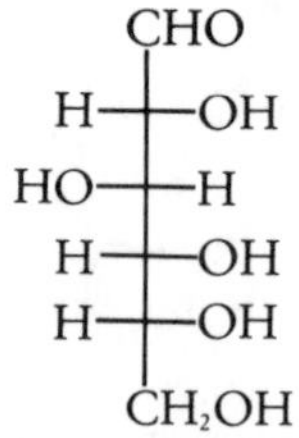

Figure 9.11   The open-chain formulation of glucose.

Ring formation is a result of intramolecular reaction between the aldehyde group and an alcoholic hydroxy group to give a hemiacetal. Two of the diastereoisomeric pyranoses are commonly found for glucose, as shown in Figure 9.12. The more stable form is β-D-glucose in which all the substituents are equatorial. In α-D-glucose only the hemiacetal hydroxy group is axial. The two forms of glucose are in equilibrium in solution but β-D-glucose predominates (about 95%). Sugar chemists call the epimeric glucoses (and other sugar epimers) *anomers*.

Poly (glucose) in one form or another is one of the most widespread chemicals in the living world. Cellulose is the major constituent of cotton and flax and is inedible. Starch, on the other hand is found extensively in food-stuffs such as rice, wheat and potatoes. The difference between starch and cellulose is simply that starch is poly (α-D-glucose) and cellulose is poly (β-D-glucose). Figure 9.13 shows the dimerization of α-D-glucose to give maltose, which can then polymerize through linkage of the equatorial OH group indicated on the left-hand glucose ring, and the axial hemiacetal OH group on the right-hand ring. Extra stability of the starch chains is gained through intramolecular hydrogen bonding as shown.

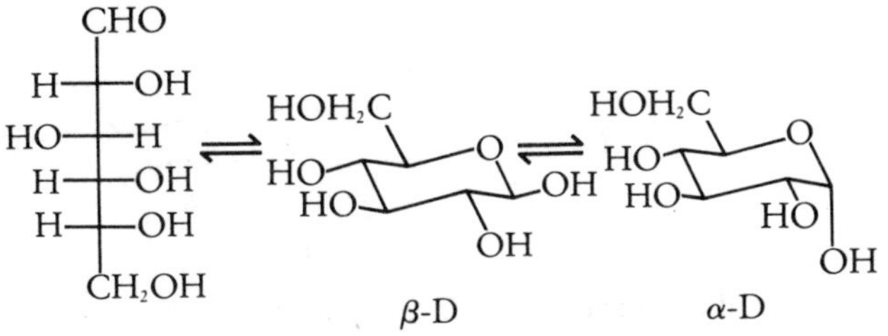

Figure 9.12 The diastereoisomeric pyranose forms of glucose, α-D-glucose and β-D-glucose.

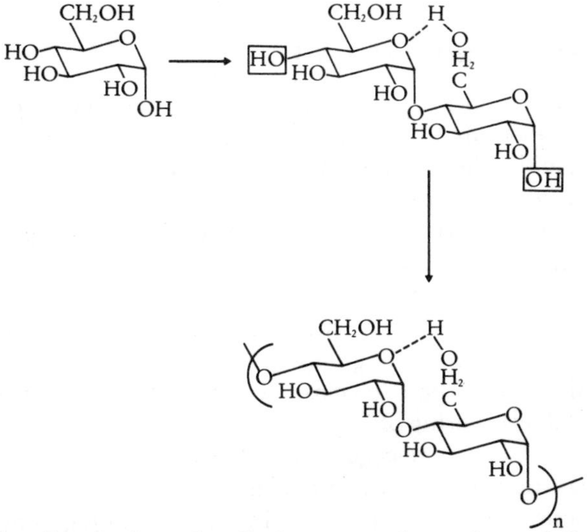

Figure 9.13 The dimerization of α-D-glucose to give maltose, followed by polymerization to starch.

β-D-glucose is similarly able to give the cellulose precursor, cellobiose, as shown in Figure 9.14 (p. 130) which can then polymerize to cellulose. There is stronger hydrogen bonding in cellulose than in starch owing to the very favourable orientation of the ring oxygen with the equatorial hydroxy group on the neighbouring glucose residue.

The difference between starch and cellulose is only one of configuration at the linkages between the monomer unit. This is an excellent example of the profound effect that configuration can have on biological properties.

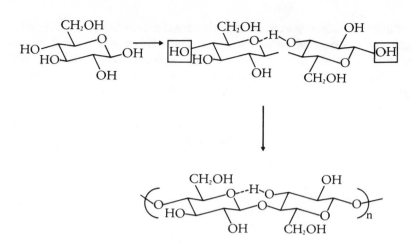

Figure 9.14 The dimerization of β-D-glucose to give cellobiose, followed by polymerization to cellulose.

9.5 Summary of Sections 9.1–9.4

1. Macromolecules are molecules with high molecular weights. Polymers are compounds with well-defined repeat units and are most often long chain molecules. Homopolymers have the same repeat unit; copolymers may have a number of different repeat units.
2. Polyethylene chains in the solid state exist in an all-antiperiplanar lowest-enthalpy conformation.
3. The tacticity of a polymer describes the configurational relationship of side chain groups. A polymer $(RCHCH_2)_n$ is isotactic if in the all-anti-periplanar form all R groups are on the same side of the molecule, syndiotactic if the R groups alternate and atactic if there is no well-defined relationship of R groups.
4. Syndiotactic polymers generally adopt an all-antiperiplanar chain structure as steric hindrance is minimized.
5. Isotactic polymers adopt a helical structure with R groups pointing away from the chain. These helices may be left- or right-handed and have a screw axis.
6. Atactic polymers have no well-defined stereochemistry.
7. Poly (amino acids) if made from amino acids of the same series are chiral. Polyglycine is the exception and adopts an antiparallel sheet structure.
8. A β-pleated sheet structure is common for poly (amino acids) in which the R groups are disposed alternately above and below the sheet. The amide bond is almost always Z and is always planar. The dihedral angles ψ and ϕ may change and affect the secondary structure. In a pleated sheet ϕ is $\sim -140°$ and ψ is $\sim +145°$ for L-amino acids.

9. Smaller values for ψ and ϕ lead to a helical structure. The right-handed helix in which R eclipses the NH group is favoured and ϕ is $\sim -47°$ and ψ $\sim -58°$ for an α-helix with 3.6 amino acid residues per turn and L-amino acids. The R groups are disposed around the outside of the helix. The left-handed helix is diastereoisomeric with the right-hand helix owing to the chirality of the amino acids.

10. Polysaccharides are polymers of the pyranose glucose ring, which may have axial (α) or equatorial hemiacetal OH groups (β).

11. Cellulose is a polymer of β-D-glucose and starch a polymer of α-D-glucose, differing only in the configuration at the hemiacetal carbon atom.

Problems and Exercises

1. If you have not already done so construct models of the helical isotactic polypropylene and poly (L-alanine). Make sure that you understand the conformational and configurational requirements for each.

2. How would you expect the Cotton effect to differ for the right- and left-handed helices of (a) polyisotactic propylene (b) poly (L-alanine). Would you expect to *observe* a Cotton effect for a solution of either poly (isotactic propylene) or poly (L-alanine)? Explain.

3. Attempt to draw extended conformations of the stereoregular polymers of

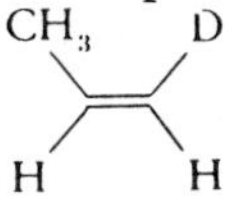

CH$_3$　　D

Are any of the chains chiral?

H　　H

PART III

The Stereochemistry of Some Chemical Reactions

Now, all the background is available to study the relationships between shape and reactivity, or form and function as biologists have it.

Following a revision chapter on chemical reactivity (Chapter 10) the subsequent chapters examine, substitution (Chapter 11), elimination and addition (Chapter 12), pericyclic reactions (Chapter 13) and finally asymmetric synthesis (Chapter 14).

The principles of chemical reactivity

10.1 Introduction

In general a study of reaction dynamics precedes a detailed study of reaction stereochemistry. We therefore assume a basic knowledge of reaction chemistry, with this chapter serving mainly as revision material. The remaining chapters in the book are devoted to the stereochemical consequences of chemical reactions, and the principles that determine them. One of the most crucial aspects of reaction stereochemistry is concerned with the planned, logical synthesis of one stereoisomer, of a particular set of stereoisomers. This assumes a major importance in the synthesis of pharmaceutical compounds and in the synthesis of relatively large amounts of compounds that can be isolated in only minute quantities from natural sources. As an example, if a desired stereoisomer has five chiral centres there will be thirty-two possible stereoisomers arising from these alone. A synthesis that produced each stereoisomer randomly would have a maximum yield of about 3% for any stereoisomer before any other factors were considered! The need for syntheses leading preferentially to one stereoisomer is obvious.

The ability to synthesize particular stereoisomers derives from an intimate knowledge of *reaction mechanisms*. A reaction mechanism is an attempt to describe, *at the molecular level,* the changes that take place during a chemical reaction.

Particular reaction mechanisms are discussed in forthcoming chapters. Here we shall review some fundamental ideas of chemical reactivity that preface a study of mechanisms. First, let us examine the different classes of chemical reaction, *substitution, addition, elimination* and *rearrangement.*

Substitution reactions may be expressed as,

$$A + BC \longrightarrow AB + C$$

for example,

$$CN^- + (CH_3)_2CHCl \rightleftharpoons (CH_3)_2CHCN + Cl^-$$

Addition reactions,

$$A + B \longrightarrow AB$$

for example,

$$Cl_2 + H_2C = CH_2 \longrightarrow ClCH_2CH_2Cl$$

Elimination reactions are the reverse of addition,

$$AB \longrightarrow A + B$$

for example,

$$C_2H_5OH \xrightarrow{\ H^+\ } CH_2 = CH_2 + H_2O$$

and finally *rearrangements,*

$$ABC \longrightarrow CAB$$

for example,

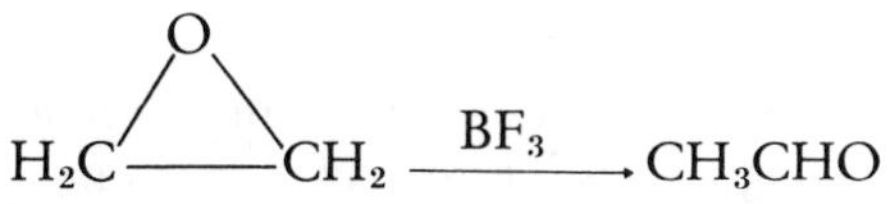

$$H_2C \overset{\displaystyle O}{\underset{\textstyle CH_2}{\triangle}} \xrightarrow{\ BF_3\ } CH_3CHO$$

Complex reactions can be a combination of substitution, addition, elimination and rearrangement reactions; but in all cases reactions may be broken down into a series of steps each involving one of these basic reactions.

10.2 The thermodynamics and kinetics of chemical reactions

10.2.1 The driving force for reactions

The first problem to be tackled in discussing chemical reactions is the question of whether they will occur at all. The *thermodynamic* requirement for a spontaneous chemical reaction is that the Gibbs free energy, $\Delta G^\ominus$, of the system should decrease continually during a reaction. Equilibrium is reached when the free energy reaches a minimum and no longer spontaneously changes. The difference in standard free energies of the reactants and products is related to the equilibrium constant, K, through the familiar expression,

$$\Delta G^\ominus = -RT \ln K, \text{where} \quad \Delta G^\ominus = \Delta H^\ominus - T\Delta S^\ominus$$

This is an expression of the universal observation that any system seeks a position of minimum energy. Rivers, for example, flow towards the seas.

For a reaction to proceed spontaneously it is not a sufficient condition for the Gibbs free energy to decrease; the reaction must also proceed at a measurable *rate*. There must therefore be a path of sufficiently low energy for the reaction to follow. Land-locked lakes cannot flow towards the sea unless a continuously falling path can be made. There are numerous chemical reactions that would occur with the liberation of free energy, if a suitable path could be found between reactants and products. All living matter is *thermodynamically unstable* with respect to reaction with oxygen. Despite sporadic accounts of spontaneous combustion of human beings, including a fictional one by Charles Dickens in *Bleak House*, we do not, fortunately, burst into flames on contact with oxygen!

Chemical systems for which spontaneous reaction is feasible but which do not react are said to be kinetically *inert*.

10.2.2 The rates of reactions

The study of reaction rates is called *kinetics*. In the study of chemical reactions a distinction is always made between kinetics and thermodynamics; how *fast* a reaction proceeds and how *far*. Note that the adjectives stable and unstable pertain to thermodynamics, and labile (reactive) and inert (unreactive) refer to kinetic properties.

The rates of chemical reactions are usually quantified by means of an experimentally determined *rate equation*, of the form,

$$\text{rate} = k[A]^a[B]^b[C]^c \ldots.$$

where k is the *rate constant* and the powers, a, b, c, etc., show the dependence on the concentration of various reacting species. The sum of the powers, a, b, c, etc. is called the *order* of the reaction. Most simple reactions are first or second order, having rate equations,

$$\text{rate} = k[A]$$

for first order reactions and,

$$\text{rate} = k[A]^2 \text{ or rate} = k[A][B]$$

for second order reactions.

Complex, multistep reactions can have complex rate equations, that vary with the concentrations of the reacting species. The kinetics of such systems are dealt with in standard texts.

We shall concentrate attention, for the moment, on reactions that take place within a single step. These are reactions that proceed from reactants to products without forming intermediate species at energy minima.

A spontaneous chemical reaction takes place with a continuous liberation of Gibbs free energy for the system. However, each combination of reactants, that proceeds to form products requires a minimum energy content, before that reaction can proceed. The combination of molecules with this energy is said to be the *activated complex*. Strictly, this term refers to the molecular aggregation and it is said to pass through the *transition state*. These two terms are often wrongly used interchangeably.

The difference in energy between the ground state energy of the reactants, and the transition state is called the *activation energy*. Now the rate equation can be used to deduce the number of each type of reactant molecule in the activated complex. The sum of the number of molecules aggregated in the activated complex is called the *molecularity* of the reaction and is rarely greater than two. In many cases the molecularity is equal to the order of reaction. However, care must be taken in the interpretation of rate equations. Consider the case of a reaction, $A + B \rightarrow C$. The rate equation describes the dependence of rate on the concentrations of A and B. If A is in large excess, perhaps as the solvent, its concentration will not change measurably during reaction and it will not appear in the rate equation, *even if the activated complex is* $[AB]^{\ddagger}$. The molecularity in this case will be two and the order one. Other means of establishing the dependence on A must therefore be found. This illustrates the need for care in the interpretation of rate equations.

10.2.3 The activation energy

In certain cases spectroscopic studies can lead to Gibbs energies of activation. For chemical reactions the measurement of the variation of the rate of reaction at a number of different temperatures allows the calculation of an activation energy E_a, from the *Arrhenius equation*

$$k = Ae^{-E_a/RT}$$

where k is the rate constant and A is a constant often called the *preexponential factor*. The activation energy, E_a, is related to the activation enthalpy $\Delta H^{\ddagger}$ by,

$$E_a = \Delta H^{\ddagger} - RT$$

Any reaction with an activation energy is faster at higher temperatures, as may be readily deduced from the Arrhenius equation.

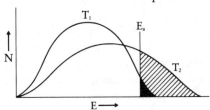

Figure 10.1 The distribution of molecular energies of a system at two different temperatures, where $T_2 > T_1$.

This is understandable in statistical terms. An ensemble of molecules, at a given temperature, do not all have the same energy. There is continual energy transfer through collisions. Figure 10.1 shows a typical distribution of energies for a collection of molecules at two temperatures where $T_2 > T_1$. The activation energy for a hypothetical reaction is marked as E_a. The number of molecules with sufficient energy for reaction is given by the shaded areas, and

137

is much greater at the higher temperature. As there are more molecules with energies sufficient for reaction to occur at the higher temperature the reaction is faster.

10.2.4 Reaction profile diagrams

Organic chemists like to represent the course of a reaction pictorially, through reaction coordinate diagrams. Three such diagrams are given in Figure 10.2 for hypothetical, simple reactions in which reactants, R, are transformed into products, P, through a single energy maximum; the transition state, $A^{\ddagger}$. The units of the 'energy' and 'reaction coordinate' axes are usually rather ill-defined. The energy scale may be marked 'free energy', 'enthalpy', 'potential energy' or simply 'energy'. The 'reaction coordinate' is usually considered to be related to changes in molecular geometry, such as bond angle or length and gives an indication of how far along the reaction path between reactants and products the molecules have travelled. Such diagrams are not precise, nor are they meant to be, they are really a device for helping us to understand the progress of a reaction in molecular terms.

It may appear that these diagrams are contradictory since we have already stated that the free energy of the *system* decreases throughout a reaction. But *reaction profile diagrams do not represent the energy of a system, but of isolated aggregates of molecules, involved in the reaction,* starting with the ground state energy of the reactants.

The interpretation of such figures can be explained by reference to Figure 10.2. The reaction represented in Figure 10.2(a) will proceed spontaneously from R to P. The activation energy is relatively small and the free energy of the products is lower than that of the reactants. By contrast, the reaction illustrated in Figure 10.2(b) has a high activation energy and is therefore very slow at normal temperatures. The reaction will not measurably proceed (in either direction) even though it is thermodynamically favoured in the forward direction. The kinetic factors dominate the thermodynamic in this example. For most simple reactions, activation energies less than about 120 kJ mol^{-1} ensure a measureable rate. The final example in Figure 10.2(c) is of a reaction in which the free energy difference between reactants and products is very small. Providing that the activation energy is sufficiently low this reaction will proceed, resulting in an equilibrium mixture of R and P. The value of the equilibrium constant for the reaction will depend only on $\Delta G^{\ominus}$, in the usual way.

The reaction profile of a particular reaction can be altered by the addition of catalysts. A catalyst is a species that can change the reaction mechanism and thereby lower the activation energy by providing an alternative, lower energy pathway. A catalyst can therefore increase the rate of a reaction but can have no effect on the equilibrium constant. At the end of a reaction a catalyst may be recovered chemically unchanged. Through the use of a suitable catalyst a reaction of type 10.2(b) can be made to proceed smoothly with a reaction

138

profile similar to that in reactions of type 10.2(a).

10.2.5 Competing reactions

The reaction profiles described in the previous section allow only for the possibility $R \rightarrow P$. A more usual reaction type encountered is one in which reactants R may give products $P1$ and/or $P2$. This is known as a *competing reaction*, and assumes particular interest where $P1$ and $P2$ are stereoisomers.

Figure 10.3 shows two different reaction profiles for $R \rightarrow P1 + P2$, through simple, one-step transformations. The first example, given in Figure 10.3(a) shows a reaction in which the activation energy for formation of products $P2$ is less than that for formation of $P1$. Similarly the products $P2$ are of lower energy than products $P1$. In this case both kinetic and thermodynamic criteria favour the formation of $P2$, which will always be produced in greater quantity than $P1$. If a catalyst can be found to reduce the activation energy for $R \rightarrow P1$ *selectively*, then the proportion of $P1$ can be increased. Temperature is also important. At very low temperatures the proportion of $P2$, formed with the lower activation energy, is greatest. As the temperature increases so the ratio of $P1 : P2$ approaches that appropriate to the thermodynamic equilibrium.

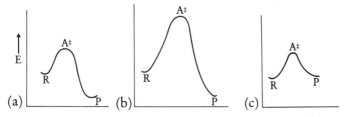

Figure 10.2 Qualitative 'reaction profile' diagrams for the conversion $R \longrightarrow P$, in a single step with no intermediate species.

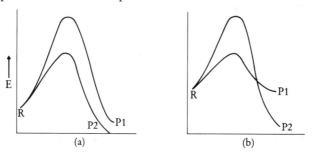

Figure 10.3 Reaction profiles for competing reactions, $R \longrightarrow P1 + P2$.

A more complex but common, example of competing reactions is given in Figure 10.3(b). The activation energy for the formation of $P1$ is lower than that for $P2$ formation, but $P2$ is favoured by thermodynamic criteria. If it is observed that $P1$ is formed in considerable excess over $P2$ then the reaction is said to be under *kinetic control*. If $P2$ is formed in large excess (or, if the free

139

energy difference between $P1$ and $P2$ is small and a thermodynamic equilibrium between $P1$ and $P2$ is achieved) then the reaction is said to be under *thermodynamic control*.

Without an intimate knowledge of the reaction mechanism and of the various values of $\triangle G^{\ominus}$ and E_a it is not possible to state whether a reaction will be under kinetic or thermodynamic control. Comparison with a closely related reaction of known outcome may help in the prediction of the course of an unknown reaction.

In general, if there is a *large* difference in the activation energy leading to the two sets of products, then kinetic control will be observed. If, on the other hand, *both reverse* reactions $P2 \rightarrow R$ and $P1 \rightarrow R$ proceed at measurable rates, under the reaction conditions then thermodynamic control is inevitable. The use of catalysts is important in these examples as the ratio of $P1/P2$ can be drastically changed by their use. Once again, the temperature used will affect the ratio of $P1/P2$, lower temperatures favouring $P1$ and higher temperatures favouring $P2$.

The product ratios for kinetic and thermodynamic control can be summarized:

for *thermodynamic control:*

$$\frac{[P1]}{[P2]} = \frac{K_1}{K_2} = e^{\frac{\triangle G^{\ominus}}{RT}}$$

and for *kinetic control:*

$$\frac{[P1]}{[P2]} = \frac{k_1}{k_2} = e^{\frac{\triangle G^{\ddagger}}{RT}}$$

10.2.6 Consecutive reactions and reaction intermediates

The reactions described above all take place with an increase in energy to a single energy maximum followed by a decrease in energy to give products. Such single step reactions are common but not exclusive. Many reactions proceed through a number of steps with several energy maxima.

A reaction $A \rightarrow B \rightarrow C$ where A, B and C are all stable, isolable compounds is a *consecutive reaction*.

A reaction $R \rightarrow I \rightarrow P$ where I is at a local energy minimum, but is not a stable compound is a special case of a consecutive reaction called a *stepwise reaction* which proceeds through a *reactive intermediate*, I. Intermediates, being at energy minima, may be observed in principle, as they have a finite concentration, by contrast to transition states which cannot be observed.

Figures 10.4(a) and 10.4(b) show two types of reaction profiles for two-step reactions with an intermediate I. The first reaction profile shows the progress of a reaction in which the transition state, $A_1^{\ddagger}$, leading to I is of higher energy than the second energy maximum, $A_2^{\ddagger}$, which lead to products. In these reactions the activation energy is the difference in energy between reactants R and the highest energy maximum. The greatest energy barrier to be surmounted in the first profile is $R \rightarrow A_1{}^{\ddagger}$. For this stepwise reaction the step that determines the overall rate of reaction is $R \rightarrow I$. This step is called the *rate determining step, r.d.s.,* or the *rate limiting step.* In other words, for this reaction the rate of reaction is equal to the rate of formation of the intermediate. No matter how fast the intermediate decomposes to products it cannot decompose faster than it is formed.

The other reaction profile, shown in Figure 10.4(b) shows an opposite behaviour, in which the breakdown of the intermediate is slower than its formation, and the rate limiting step is $I \rightarrow P$. The overall activation energy as derived from the Arrhenius equation is the difference in energy between R and $A_2{}^{\ddagger}$.

The analysis of competing reactions proceeding through several steps is more complex than with simple single-step reactions but follows the same principles. In a reaction $R \rightarrow P1+P1$ in which $P1$ and $P2$ are formed from a common intermediate it is relatively difficult to vary the ratio of $P1$ to $P2$.

A more common example, studied in Chapter 12, has different, diastereoisomeric, intermediates leading to different products. Catalysts, that lower the energy of one particular intermediate can affect the product ratio considerably. Temperature and solvent changes may also assume importance.

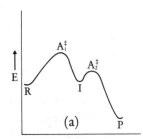
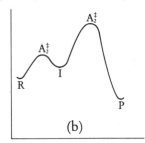

Figure 10.4   Reaction profiles for reactions R ——→P that proceed through an intermediate I.

10.3 Selectivity and specificity in chemical reactions

It is useful to be able to express the precision of chemical reactions with respect to attack at a particular position in a molecule and/or production of constitutional isomers, diastereoisomers or enantiomers.

The generation of centres of chirality from achiral centres is covered in Chapter 14 and selectivity of enantiomer formation is discussed there. For the present we are only concerned with reactions that have an element of symmetry, so that enantiomers are formed in equal quantities.

Now we present some definitions of selectivity and specifity with suitable examples. *Chemoselectivity.* A reagent that will transform one particular functional group is said to be *chemoselective*. A particular example is the addition of sodium bisulphite, $NaHSO_3$, to aldehydes but not ketones,

$$NaHSO_3 + OHC\text{\wwc}C{\overset{O}{\underset{R}{\diagdown}}} \rightarrow Na\overset{+}{O}{}^{-}\overset{\overset{H}{|}}{\underset{\underset{OSO_2H}{|}}{C}}\text{\wwc}C{\overset{O}{\underset{R}{\diagup}}}$$

Regioselectivity. A reaction is said to be regioselective if one of a possible set of constitutional isomers is produced in considerable excess. Examples are,

$$\underset{H}{\overset{Ph}{\diagdown}}C=C\underset{H}{\overset{H}{\diagup}} + HCl \rightarrow \underset{H}{\overset{Ph}{\diagdown}}\overset{\overset{Cl}{|}}{C}-CH_3$$

for an additional reaction, in which the reagent, HCl, adds to the double bond with one particular orientation. Elimination reactions can also be regio-selective,

$$\underset{CH_3}{\overset{CH_3}{\diagdown}}\overset{\overset{Cl}{|}}{C}-CH_2CH_3 \quad\underset{}{\overset{base}{\rightarrow}}\quad \underset{CH_3}{\overset{CH_3}{\diagdown}}C=C\underset{H}{\overset{CH_3}{\diagup}}$$

More important for stereochemical studies are *stereoselective* and *stereo-specific* reactions. These terms are frequently used in a rather imprecise way. We present here the definitions of Eliel which are those most widely accepted by chemists. These terms should only be applied to reactions with an element of symmetry.

A stereoselective synthesis is one producing one diastereoisomer (or enantiomeric pair of diastereoisomers) *in excess over another*. The degree of selectivity is often described qualitatively (highly, slightly, etc.) or quantitatively in terms of either a ratio or percentage excess of one diastereoisomer. An example of a stereoselective reaction is the reduction of but-2-yne with hydrogen in the presence of a Lindlar catalyst.

$$CH_3-C\equiv C-CH_3 + H_2 \xrightarrow[\text{catalyst}]{\text{Lindlar}} \underset{H}{\overset{CH_3}{\diagdown}}C=C\underset{H}{\overset{CH_3}{\diagup}} \qquad (10.1)$$

Only *cis-(Z)*-but-2-ene is formed in this reaction with no observable amount of the *E(trans)* isomer. In this case the selectivity arises from the nature of the reaction and not from any stereochemical feature of the reactants.

Another stereoselective reaction is the formation of maleic anhydride from both maleic *and* fumaric acid,

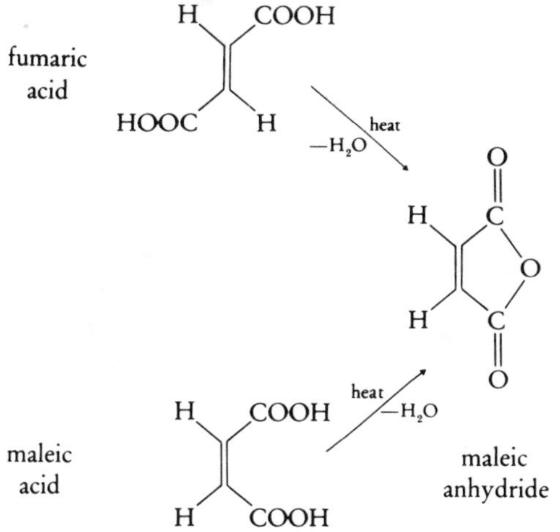

fumaric
acid

$$H \quad COOH$$
$$HOOC \quad H$$

heat
$-H_2O$

maleic
acid

$$H \quad COOH$$
$$H \quad COOH$$

heat
$-H_2O$

maleic
anhydride

(10.2)

In this case the selectivity arises from the geometric requirements of the product; a *trans* anhydride cannot be formed.

Selectivity can arise from kinetic, thermodynamic or stereochemical constraints.

A stereospecific reaction is one in which stereochemically different reactants lead to stereochemically different products.

One example of a stereospecific reaction is the addition of bromine to *cis* and *trans*-but-2-ene,

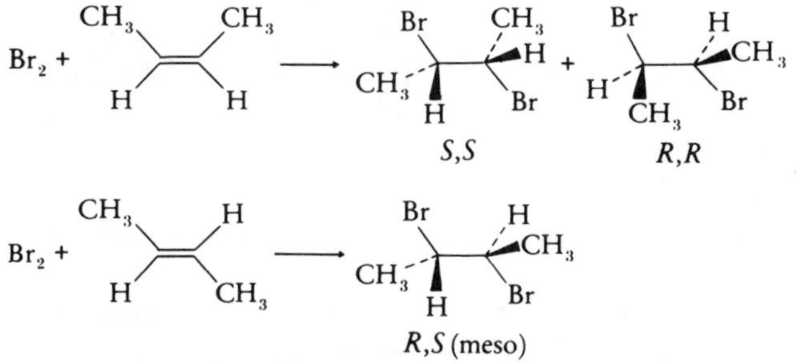

$$Br_2 +$$

S,S

R,R

$$Br_2 +$$

R,S (meso)

This reaction is studied in detail in Chapter 12. It should be noted that the above reactions are both stereoselective as well as being stereospecific. *All stereospecific reactions are necessarily stereoselective but stereoselective reactions are not necessarily stereospecific.* The two examples of stereoselectivity given above (equations 1 and 2) are *not* stereospecific.

Once again the degree of stereospecificity can be described qualitatively or quantitatively.

In the following chapters the principles and methods described in this chapter will be used to examine the stereochemistry of chemical reactions. The first reaction type to be examined is *substitution*.

CHAPTER 11

The stereochemistry of substitution reactions

11.1 The effect of substitution on configuration

The predominant topic of this chapter is substitution reactions of the type,

abcCX → abcCY

in which one group, X, is substituted by another, Y, and the rest of the molecule remains constitutionally unchanged.

The stereochemical changes of interest are those relating to the configuration of the abcC-fragment of the molecule. Such changes can only be observed when the configuration of the substituent-bearing carbon atom is identifiable. First, let us consider the stereochemical possibilities for substitution of one enantiomer of a chiral molecule abcCX. Three extreme effects on configuration have been observed.

1. *Inversion of configuration.* If a substrate, abcCX, has the configuration shown in Figure 11.1(a) and the product has the configuration given in Figure 11.1(b), then the reaction is said to occur with inversion of configuration. This inversion, called a *Walden inversion* is perhaps a molecular equivalent of an umbrella turning inside out in a high wind.

It should be noted that the term inversion refers to the local symmetry of the abcC fragment and does not imply a change of configuration from $R \rightarrow S$ or $S \rightarrow R$ between the substrate and product. It is quite possible that the configuration of abcC— could invert during a substitution with the product having the same configuration label as the substrate. The labels R and S depend on our imposed priority system for ligands rather than any intrinsic symmetry of the molecule.

2. *Racemization.* When a substrate such as that shown in Figure 11.1(a), or its enantiomer, reacts to give a racemic mixture of products, (11.1(b) and 11.1(c))

then the reaction occurs with racemization. In the extreme case, exactly equimolar quantities of enantiomeric products are obtained, regardless of the configuration of the starting material.

3. *Retention of Configuration.* The third stereochemical possibility for a substitution reaction is represented by the transformation in Figure 11.1(a) to 11.1(c) in which the configuration about the central carbon atom is maintained during the reaction.

Each of these stereochemical pathways will now be examined separately, with the emphasis on the mechanism and conditions favouring each route.

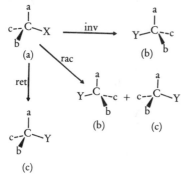

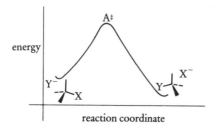

Figure 11.1   Stereochemical possibilities of substitution; inversion, racemization and retention of configuration.

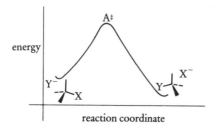

Figure 11.2   The reaction profile for an S_N2 reaction.

11.2 The mechanism of inversion of configuration

Experiments on nucleophilic substitutions that take place with inversion of configuration show that this stereochemistry is associated with a bimolecular reaction. The rate equation is,

rate = k[substrate] [nucleophile]

Reactions following these kinetics and taking place with inversion of configuration are called S_N2 reactions, for *substitutions, nucleophilic and bimolecular.*

There are no intermediate species in an S_N2 reaction and a typical reaction profile is shown in Figure 11.2. The activated complex has an approximately

145

trigonal bipyramidal geometry illustrated in Figure 11.3. The entering group, Y, and the leaving group, X, occupy the axial positions. The a, b and c ligands are equatorial and approximately coplanar in a plane containing the central carbon atom. The attacking nucleophile must attack along the C—X internuclear axis, as shown in Figure 11.3. As the bond to Y is being formed, the bond to X is being broken, synchronously.

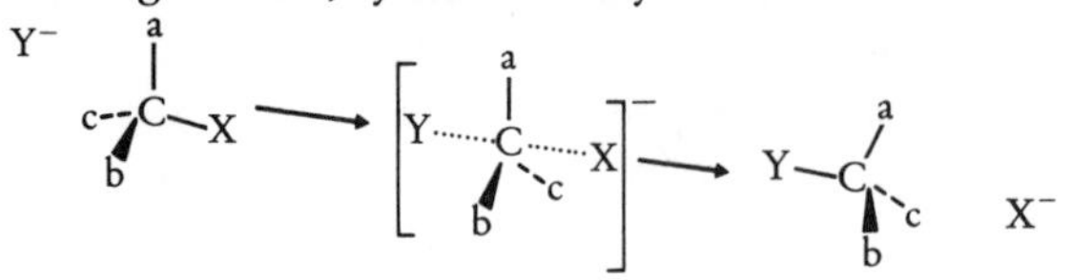

Figure 11.3 The trigonal bipyramidal geometry of the activated complex in S_N2 substitutions.

This attack from the back is the characteristic feature of S_N2 reactions. The exact geometry of the transition state depends on a number of factors, and the degree of bond breaking can be either greater or less than the degree of bond making. Only in the case where X = Y is the transition state necessarily completely symmetrical (at least C_s) with the C—X and C—Y bonds having the same length and strength.

Bimolecular, nucleophilic substitutions in which no intermediate is formed, are always accompanied by exclusive inversion of configuration. It is reasonable to seek a molecular explanation for this phenomenon. At first sight a reasonable, alternative path seems to be attack by Y on the same side of the molecule as X, giving *retention* of configuration. One explanation for the exclusive rear-side attack, that frequently appears in texts, is that dipole–dipole repulsions keep X and Y apart. This may be reasonable for,

$$I^- + C_2H_5(CH_3)CHBr \rightleftharpoons IHC(CH_3)C_2H_5 + Br^-$$

but dipole–dipole *attractions* might be expected for,

$$C_2H_5(CH_3)CH\overset{+}{N}(CH_3)_3 + CH_3CO\overset{-}{O} \rightarrow$$
$$CH_3COOCH(CH_3)C_2H_5 + N(CH_3)_3$$

which, in fact, follows all other S_N2 reactions in proceeding with *inversion* of configuration.

A qualitative estimate of the relative energies of the bimolecular transition states leading to inversion and retention can be made by examining the orbital overlap in each case. These are shown in Figure 11.4 and calculations bear out the simple picture presented here, that the transition state leading to inversion is of much lower energy than that for retention. Although the transition state for retention of configuration is apparently so high in energy that reaction along that path cannot proceed for carbon, retention is known for chiral silicon compounds. This is a different case, however, as stable five coordinate silane anions R_5Si^- have been *isolated*.

The characterization of inversion of configuration as a stereochemical feature of S_N2 reactions was not always as easy as it can be now, with the availability of X-ray and chiroptical methods. Chemical correlations were used and in some cases a cycle of reactions was set up in which one step had unknown stereochemistry. By this means the stereochemistry of a reaction could be determined. One well-known correlation sequence is given in Figure 11.5. In three steps an alcohol of one configuration is converted relatively cleanly into its enantiomer. The configuration is unchanged in step 1, as the only bond broken in the chiral alcohol is the O—H bond, which does not affect the configuration. Similarly in step 3 wide experience has shown that in

the cleavage of esters, $\overset{\overset{\text{O}}{\parallel}}{\text{RCOR}'}$, the R'—O bond remains intact and the C—OR bond is broken. Therefore there is no change in configuration at step 3. As the 1-phenyl-2-propanol at the beginning of the reaction had α_D of $+33.0$ and $-32.2°$ at the end of the sequence, practically quantitative inversion of configuration must take place at the nucleophilic substitution in step 2.

Figure 11.4   Qualitative comparison of the orbital overlap in the inversion (a) and retention; (b) transition states for S_N2 substitution.

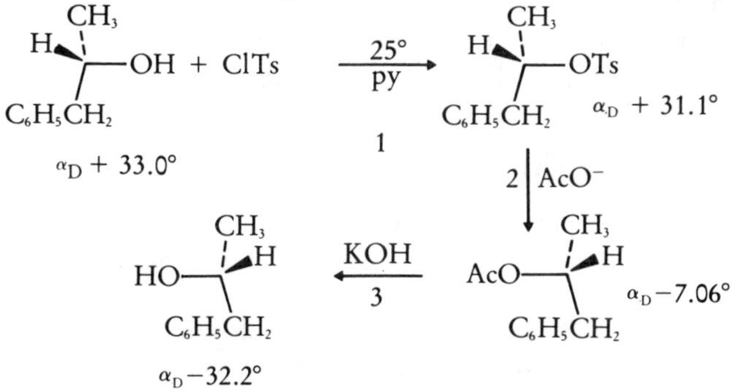

$$
\begin{array}{c}
\text{CH}_3 \\
\text{H}_{\diagdown} \overset{|}{} \\
\diagup \text{—OH} + \text{ClTs} \\
\text{C}_6\text{H}_5\text{CH}_2
\end{array}
\xrightarrow[\text{py}]{25°}
\begin{array}{c}
\text{CH}_3 \\
\text{H}_{\diagdown} \overset{|}{} \\
\diagup \text{—OTs} \qquad \alpha_D + 31.1° \\
\text{C}_6\text{H}_5\text{CH}_2
\end{array}
$$

1

$$\alpha_D + 33.0°$$

2 | AcO⁻

$$
\begin{array}{c}
\text{CH}_3 \\
\overset{|}{} \diagdown \text{H} \\
\text{HO—}\diagup \\
\text{C}_6\text{H}_5\text{CH}_2
\end{array}
\xleftarrow[3]{\text{KOH}}
\begin{array}{c}
\text{CH}_3 \\
\overset{|}{} \diagdown \text{H} \\
\text{AcO—}\diagup \qquad \alpha_D - 7.06° \\
\text{C}_6\text{H}_5\text{CH}_2
\end{array}
$$

$$\alpha_D - 32.2°$$

Figure 11.5   The demonstration of inversion of configuration in an S_N2 substitution.

Inversion of configuration has also been demonstrated by the substitution of racemic *trans*-1,2-disubstituted cyclohexanes. These compounds are chiral, but do not need to be resolved, as substitution of one ligand can give one pair of diastereoisomers (*trans*) by retention and another (*cis*) by inversion. As diastereoisomers can readily be distinguished physically and chemically there is no need for a resolution. The scheme in Figure 11.6 on p. 148 shows that the observed inversion of configuration, results in *cis* isomers being formed.

The need for attack at the rear is illustrated by the unreactivity towards S_N2 substitution of bridgehead compounds, such as the one shown in Figure 11.7.

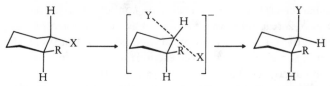

Figure 11.6   The demonstration of inversion of configuration in an S_N2 substitution of *trans*-1,2-disubstituted cyclohexane.

Figure 11.7   Bridgehead substituted cage compounds are unreactive towards S_N2 substitution.

The cage-like ring structure prevents the access of the incoming nucleophile at the rear, and consequently S_N2 reaction is too slow to measure.

Now that we have seen something of the nature of, and requirements for, inversion of configuration we can investigate some of the factors that promote S_N2 reactions.

In view of the need for attack from the back, it is not surprising that S_N2 reactions are favoured at relatively unhindered carbon atoms. For substituted alkyl compounds the ease of S_N2 decreases,

CH_3— > primary alkyl > secondary alkyl >> tertiary alkyl

This order exactly parallels the steric hindrance at the central carbon atom. At present there is no evidence for an S_N2 reaction at a hindered, tertiary carbon atom.

The rate of an S_N2 reaction is dependent on the concentration of both nucleophile *and* substrate. A high nucleophile concentration therefore increases the rate of an S_N2 reaction. The *nature* of the nucleophile is also important; some nucleophiles are much more effective than others in S_N2 reactions. Good nucleophiles are said to have high *nucleophilicity*, which, being concerned with rates, is a kinetic property. It is not a simple matter to rationalize the nucleophilicities of a series of nucleophiles. Polarizability, base strength and other factors all seem to be involved. However the order is well known and nucleophilicity decreases,

$C_6H_5S^-$ > CN^- $\simeq$ I^- > HO^- $\simeq$ RO^- > Br^- > C_6H_5O > Cl^- > CH_3COO^- >> NO_2^- >> ClO_4^-

The leaving group in S_N2 reaction is rather less important, except in the case of HO^- and F^-, which are extremely poor leaving groups. The hydroxyl group

148

can be made into a good leaving group by protonation, so that neutral water leaves the molecule. Another method of making a good leaving group from a hydroxyl group is to convert O—H into $O—SO_2C_6H_4CH_3—p$, the tosyl group, which is also a very effective leaving group.

11.3 The mechanisms of racemization

Racemization is almost always associated with a first order reaction in which,

$$\text{Rate} = k[\text{substrate}]$$

Reactions taking place with racemization and following first order kinetics are called S_N1 rections, for *substitutions, nucleophilic and unimolecular.*

Equal amounts of enantiomeric products are formed during the racemization reaction of a chiral substrate. This randomization of chirality implies that somewhere along the reaction path is a step in which chirality is lost, followed by a step during which either enantiomer is formed with equal probability. By contrast to S_N2 reactions, S_N1 reactions are stepwise, proceeding through a *carbonium ion* (now frequently called a carbenium ion) *intermediate*. A typical S_N1 reaction profile is illustrated in Figure 11.8. As the reaction is unimolecular, the rate determining step must be the first step, which is in this case the reversible ionization of the substrate,

$$\text{abcC—X} \overset{\text{slow}}{\rightleftharpoons} \text{abcC+X}^-$$

This step produces the carbonium ion intermediate, which is very short-lived as it can either reform the substrate, or react with the nucleophile, Y^-, to give product,

$$\text{abcC}^+ + Y^- \overset{\text{fast}}{\longrightarrow} \text{abcC—Y}$$

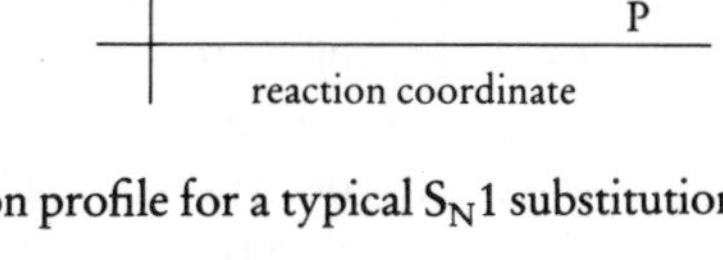

Figure 11.8   The reaction profile for a typical S_N1 substitution.

The chirality of the substrate is lost when the intermediate is formed, as carbonium ions are planar, with bond angles of 120° (shown in Figure 11.9, p. 150). As the free carbonium ion is completely planar, (abcC^+ as point group C_s,) the nucleophile can in this extreme case attack at either side with

149

equal probability. Equal amounts of enantiomers are therefore formed with racemization.

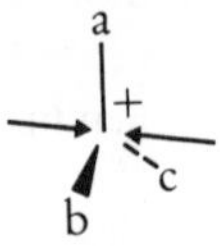

Figure 11.9 The planar geometry of carbonium ion intermediate. The arrows indicate that attack of a nucleophile is equally likely from either side.

Starting with an optically active substrate it is easy to establish racemization by monitoring the loss of optical activity.

Although each S_N2 reaction is always accompanied by inversion of configuration the stereochemical consequences of an S_N1 reaction are not always so straightforward as complete racemization. In many cases partial inversion of configuration is observed. The usual explanation for this is that the outgoing group shields one side of the molecule from attack by incoming nucleophiles. The extent of inversion, if this is reasonable, should be dependent on the carbonium ion reactivity. The most reactive carbonium ions will be attacked very quickly while the leaving group is still very close to the molecule, and there will be significant inversion. Experiments have confirmed this and also shown that the least reactive carbonium ions lead to the greatest amount of racemization.

There is overwhelming evidence that carbonium ions are planar. Pyramidal carbonium ions are of prohibitively high energy. The preference for planarity is expressed in the very slow solvolysis of bridgehead halides, compared to their open chain analogues. For example, for a silver-ion catalysed hydrolysis the following reactivity is observed,

Relative rate ≃ 110   1

A model of the cyclic halide is easy to make, but it is almost impossible to construct a model in which the C—Cl group is replaced by a trigonal planar centre, exemplifying the high energy of the carbonium ion.

The trityl, or triphenylmethyl carbonium ion, Ph_3C^+, is one of the most stable and long-lived of known carbonium ions. Ph_3CCl readily solvolyses by an S_N1 mechanism. The stability derives from delocalization of electrons from the benzene rings into the vacant p orbital. This requires the p orbitals making up the benzene aromatic system, and the vacant p orbital to be coplanar. The cyclic analogue of Ph_3CCl, shown in Figure 11.10, will not solvolyse by an S_N1 mechanism because a planar carbonium ion suitable for electron delocalization cannot be made. This geometric limitation can be

150

demonstrated by making a model of the system shown in Figure 11.10 and replacing the C—Cl bridgehead by a trigonal bipyramidal centre. The model will show that the aromatic π-systems and the vacant p orbital are at 90° and therefore non-overlapping.

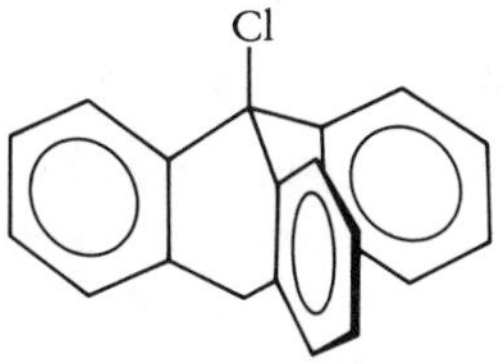

Figure 11.10   Unlike Ph$_3$CCl this compound does not form carbonium ions as delocalization of charge is geometrically impossible.

In most examples of S_N1 reactions the steric effect operates in the opposite direction to that for S_N2 reactions. Highly hindered substrates solvolyse by an S_N1 mechanism more rapidly than less hindered analogues.

Dimethylneopentyl chloride solvolyses some six hundred times more rapidly than *tertiary* butyl chloride,

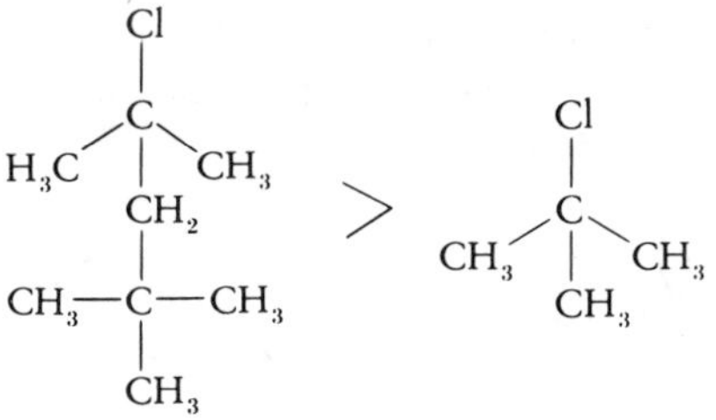

The accelerated rate for the former is most probably due to a release of steric strain on widening the bond angles to 120° in the intermediate. The general order for S_N1 solvolysis is,

tertiary > secondary > primary > CH$_3$—X

which is also the order of decreasing steric crowding. The relief of steric strain at the intermediate is thought to be one factor contributing to the above order. Others are the electron releasing effect of alkyl groups in stabilizing the carbonium ions, and a hyperconjugative effect.

The rates of S_N1 reactions of the type,

$$RX \rightleftharpoons R^+ + X^- \rightarrow RY$$

are particularly susceptible to solvent changes.

For these reactions, the transition state resembles the intermediate in bearing a much greater charge than the substrate. Ionizing solvents, of high

151

dielectric constant, strongly favour these S_N1 reactions, by stabilizing the intermediate relative to the substrate.

There is one special case where racemization is associated with an S_N2 reaction and that is reactions in which entering and leaving groups are identical,

$$X^- + abcCX \rightleftharpoons XCabc + X^-$$

Each particular substitution is accompanied by an inversion of configuration. After a finite time the mixture contains equal amounts of each enantiomer and optical activity is lost. This reaction type has been extensively used in mechanistic studies. It has limited practical application.

11.4 Mechanisms of retention of configuration

There is no straightforward single-step nucleophilic substitution path that can lead to retention of configuration. There are, however, three distinct routes by which configuration can be retained during substitution reactions.

The first is a bimolecular route called S_Ni for *substitution, nucleophilic and intramolecular*. S_Ni reactions are particularly common for the halogenation of alcohols by certain reagents. Table 11.1 gives the stereochemistry of chlorination of alcohols by different reagents, under a variety of conditions. S_Ni reactions usually occur in several steps; a typical example is,

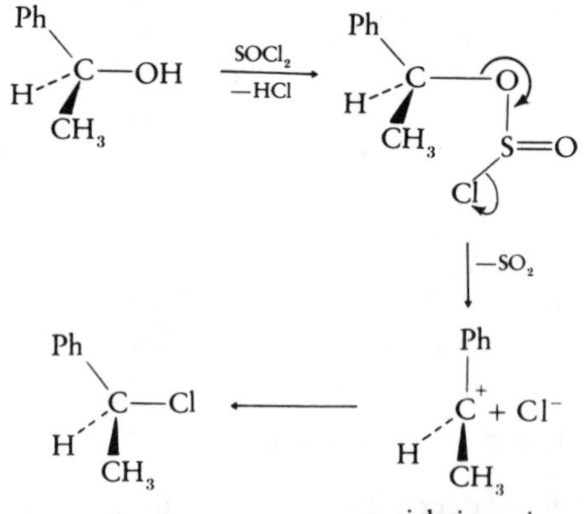

tight ion pair

The first step has no effect on the configuration as it is only a replacement of the hydroxyl hydrogen atom by the SOCl group. The subsequent step is the crucial step in which sulphur dioxide is eliminated to form an ion pair supposed to be surrounded by a solvent cage. The ion pair rapidly collapses to form a covalent chloride and the relative orientation of the ions ensures that configuration of the original alcohol is retained.

It is very likely that the carbonium ion, even with such a fleeting existence, is relatively planar before ion combination. The evidence supporting this

152

assertion is that bridgehead alcohols do not undergo S_Ni reactions whereas these open chain analogues do. Again the difficulty of attaining planarity is used as an explanation.

Table 11.1. The stereochemistry of the halogenation of alcohols, ROH → RX

Reagent	Catalyst	Solvent	Stereochemistry
PCl_3	—	any	inversion
CCl_4	PPh_3	any	retention
Br_2	PPh_3	any	inversion
$COCl_2$	—	any	retention
$SOCl_2$	—	relatively non polar	retention
$SOCl_2$	pyridine	relatively non polar	inversion
$SOCl_2$	—	DMF	inversion
$POCl_3$	pyridine	any	inversion

In a strongly ionizing solvent, such as dimethyl formanide (DMF), or in the presence of a strong base such as pyridine, the reaction above can proceed with *inversion* of configuration. The HCl formed in the first step ionizes to give chloride ions which are strongly nucleophilic, and can attack from the back, in a normal S_N2 manner, to displace the SO_2Cl^- ion with inversion of configuration.

Another route by which retention of configuration may be observed is through two successive inversions of configuration. This usually happens when there is another functional group in the substrate that can play a transient part in the reaction. Assisted reaction by a more-or-less remote functional group is called the *neighbouring group effect*. The following pair of reactions illustrate this effect. The reaction,

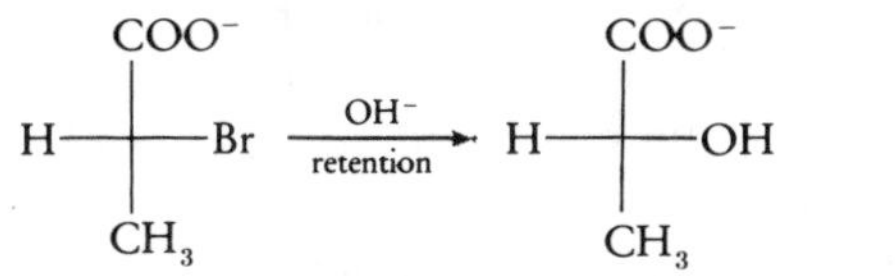

is a normal S_N2 reaction with inversion of configuration, but the superficially similar reaction,

153

proceeds with *retention* of configuration. The suggested mechanism involves two inversions, with *neighbouring group participation*,

The highly strained intermediate lactone is formed by an *intramolecular* displacement with inversion of configuration. This very reactive species is then rapidly ring-opened by hydroxyl ion, again with inversion of configuration. The overall result is therefore retention of configuration.

The final retentive mechanism is the S_E2 mechanism, for *substitution, electrophilic, bimolecular*. This mechanism is the electrophilic analogue of S_N2, but proceeds with opposite stereochemistry. This mechanism is confined to organometallic derivatives of organic compounds. As far as stereochemical usefulness is concerned the S_E2 mechanism is limited by the inability to prepare many chiral organometallic reagents. Organolithium and Grignard reagents racemize too rapidly to be useful. Only the rather toxic organomercurials are stereochemically inert. Bromination of mercurials has been shown to occur with retention of configuration,

Although there is still controversy surrounding this mechanism, it does seem that in *electrophilic* substitution the concerted path with retention of configuration is probable.

11.5 S_N2 versus S_N2

The two mechanisms labelled S_N1 and S_N2 are the extremes. In the discussion of S_N1 reactions we described examples where the nucleophile was coordinating to the central carbon atom as the leaving group was still loosely associated with the molecule. It is hard to distinguish this from an S_N2 reaction. Most chemists now accept that a continuum of mechanisms is possible, ranging between the entirely *associative* mechanism S_N2 and the *dissociative* S_N1.

There are also many known examples of competition between S_N1 and S_N2 in the same reaction. As S_N1 and S_N2 have different stereochemical consequences there is frequently good reason to prefer one mechanism rather than the other. The application of a knowledge of the mechanisms of substitution reactions often enables the conditions to be changed to be most favourable for the appropriate stereochemistry.

The solvent, substrate structure, nucleophile and, to a lesser extent, the leaving group and reaction temperature all contribute to determining the mechanism. We shall briefly examine the effect of some of these parameters.

Solvent. A reaction is favoured by an ionizing solvent (high dielectric constant) if the intermediate or transition state is more polar than the substrate. Consider the two mechanisms,

$$R-X+Y^- \rightleftharpoons R^+ \rightleftharpoons RY \qquad -S_N1$$
$$\text{or } R-X+Y^- \rightleftharpoons RY+X^- \qquad -S_N2$$

Ionizing solvents will only accelerate the rate of an S_N1 reaction of this type, while the S_N2 rate will be relatively solvent independent.

On the other hand, quaternization of amines by an S_N2 mechanism involves considerable charge separation in the transition state, as shown in Figure 11.11, and is therefore accelerated by polar solvents. The S_N1 quaternization reaction is also favoured by polar solvents.

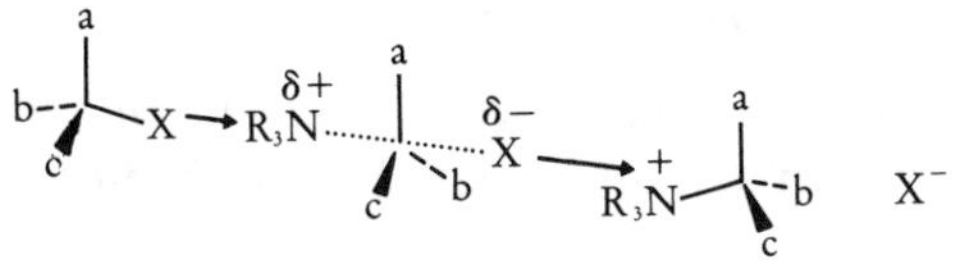

Figure 11.11   The transition state for amine quaternization.

Substrate structure. The variation of mechanism with structure can be expressed,

$$S_N2 \longleftarrow$$
$$CH_3X : \text{primary} : \text{secondary} : \text{tertiary}$$
$$\longrightarrow S_N1$$

This ordering is primarily steric in origin.

The neighbouring group effect can have a particularly dramatic effect, especially on S_N1 reactions. The temporary introduction of a neighbouring group could change, not only the mechanism from S_N2 to S_N1 but also the stereochemistry from inversion to retention. Sulphur-containing groups are particularly effective neighbouring groups.

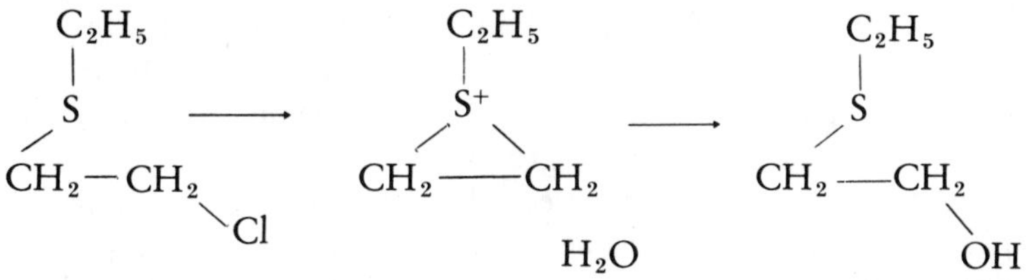

This S_N1 of 1-chloro-2-thioethylethane reaction is some ten thousand times faster than with the ether analogue and S_N1 solvolysis of chloroethane does not occur at a measurable rate.

Nucleophile. While the nature of the nucleophile will not alter the rate of an S_N1 reaction, an S_N2 reaction is dramatically altered by the nucleophile. High concentrations of a good nucleophile will favour S_N2. The following reactions give an idea of the importance of nucleophile and solvent,

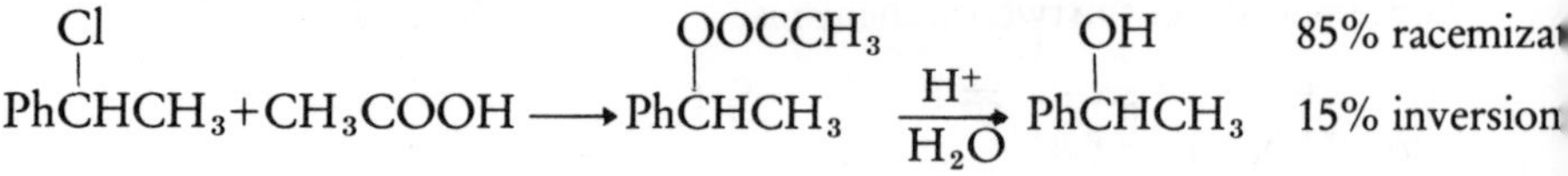

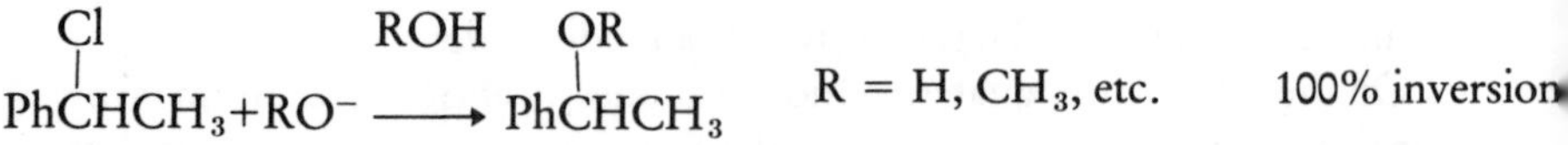

$$\underset{\text{PhCHCH}_3}{\overset{\text{Cl}}{|}} + \text{CH}_3\text{COOH} \longrightarrow \underset{\text{PhCHCH}_3}{\overset{\text{OOCCH}_3}{|}} \xrightarrow[\text{H}_2\text{O}]{\text{H}^+} \underset{\text{PhCHCH}_3}{\overset{\text{OH}}{|}} \quad \begin{array}{l} 85\% \text{ racemizat} \\ 15\% \text{ inversion} \end{array}$$

$$\underset{\text{PhCHCH}_3}{\overset{\text{Cl}}{|}} + \text{RO}^- \longrightarrow \underset{\text{PhCHCH}_3}{\overset{\text{OR}}{|}} \quad R = H, CH_3, \text{etc.} \quad 100\% \text{ inversion}$$

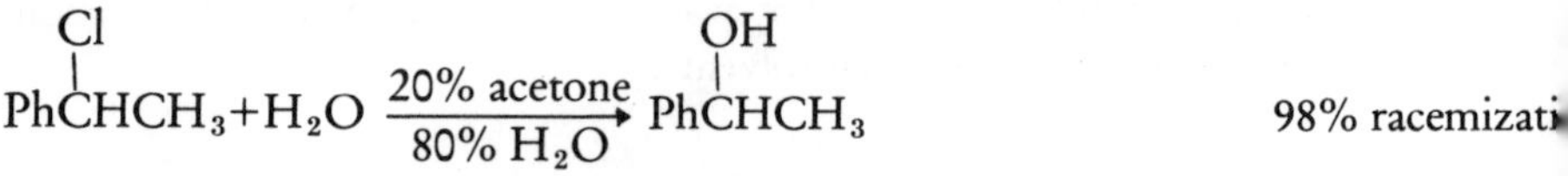

$$\underset{\text{PhCHCH}_3}{\overset{\text{Cl}}{|}} + \text{H}_2\text{O} \xrightarrow[\text{80\% H}_2\text{O}]{\text{20\% acetone}} \underset{\text{PhCHCH}_3}{\overset{\text{OH}}{|}} \quad 98\% \text{ racemizat}$$

In many cases, *elimination* reactions can compete with substitution. Elimination reactions are part of the subject of the next chapter.

11.6 Summary of Sections 11.1–11.5

1. Inversion of configuration is usually associated with S_N2 reactions, where rate $= k$[substrate][nucleophile].
2. Racemization reactions are usually S_N1, where rate $= k$[substrate].
3. Retention can be either S_Ni, through decomposition of a cyclic intermediate to give an ion pair, or two successive inversions also give retention, often with neighbouring group participation. S_E2 reactions, particularly of organomercurials occur with front-side attack to give a concerted, bimolecular, retentive mechanism.
4. S_N2 reactions with Walden inversion have no intermediates and are favoured by unhindered substrates, for S_N2 CH_3—X > 1° > 2° >> 3°. Attack by nucleophile is at the rear.
5. Good nucleophiles, $C_6H_5S^-$, CN^-, I^- favour S_N2.
6. S_N1 reactions proceed through a carbonium ion intermediate which is planar. Attack can be at either side of the carbonium ion. Inversion sometimes accompanies racemization as access by the nucleophile to one side may be restricted by the leaving group.
7. S_N1 reactions are favoured by hindered substrates, steric relief on moving to the planar intermediate is possible. For S_N1 3° > 2° > 1° >> CH_3—X.
8. The balance between S_N1 and S_N2 can be changed by variations in solvent, substrate structure, nucleophile, temperature and leaving group.

Problems and Exercises

1. This problem refers to the solvolysis of I and II by ethanol.

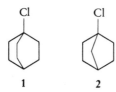

1   2

Account for the following observations:
(a) The rate of solvolysis of both **1** and **2** is unchanged by the addition of ethoxide ion.
(b) **1** solvolyses faster than **2**.
(c) The stereochemistry of substitution is exclusively retention.

2. *trans*-Br ⟨COO⁻⟩ spontaneously eliminates bromide ion in a unimolecular reaction. The *cis*-isomer is inert. Use models to help you explain these observations. What is the product of bromide elimination from the *trans* isomer?

3. R-2-chlorobutane reacts very slowly with sodium hydroxide to give S-2-butanol. A trace of iodide ion catalyses the reaction which is rapid at room temperature. Explain the role of the iodide ion and give the configuration of the product alcohol under iodide catalysis conditions.

4. Starting with a single enantiomer suggest reactions (and solvents) that could produce:
(a) $R—CH_3(C_3H_7)CHCl$ by an inversion mechanism.
(b) $R—CH_3(C_3H_7)CHCl$ by a retention mechanism
(c) $R+S—CH_3(C_3H_7)CHOTs$
(d) $R—CH_3(C_3H_7)CHOTs$.

CHAPTER 12

The stereochemistry of elimination and addition reactions

12.1 Introduction

Eliminations can take place by a number of different mechanisms to give a variety of products. We shall concentrate on one main group of alkene-producing reactions; the *1,2* or *β-eliminations,*

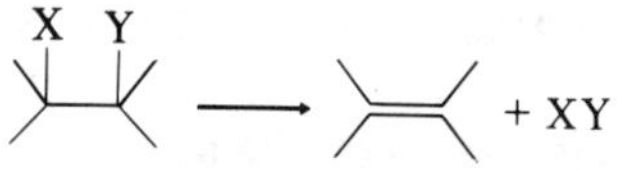

There are a number of different possible combinations of X and Y; sometimes both are halogens, but more commonly one is H and the other halogen, OR group or charged group such as $-\overset{+}{N}R_3$ or $-\overset{+}{S}R_2$.

Elimination reactions provide a general route to alkenes, and reactions are sought where one diastereomer can be produced selectively.

An added complication for both elimination and substitution reactions is that there is almost always competition between these two classes and mixtures of product types may be obtained. A study of mechanisms, as outlined below, shows how this competition arises and we end the Section on eliminations with a discussion of the factors affecting the substitution to elimination ratio.

The stereochemistry of elimination reactions parallels that of substitution reactions in its variety.

Two types of *stereospecific* eliminations are known. One is the *anti*-eliminations in which the ligands eliminated are in an antiperiplanar conformation,

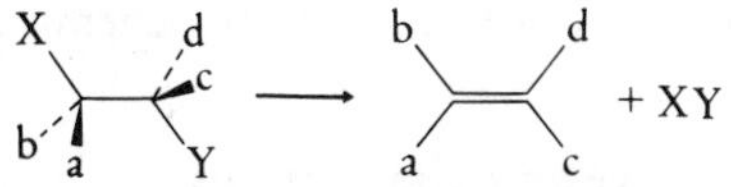

Anti-eliminations are usually base or nucleophile assisted reactions.

The other distinct, stereospecific elimination is a *syn*-elimination. Once again the four atoms involved in the elimination are coplanar but are synperiplanar in this case,

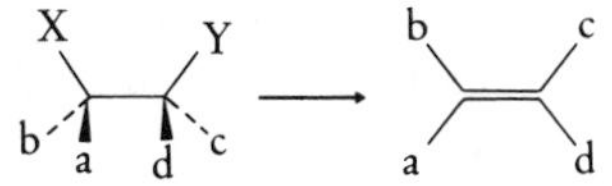

Most *syn*-eliminations are unimolecular and unaffected by added base or nucleophiles.

As both of these reactions are stereospecific, diastereoisomeric substrates lead to diastereoisomeric alkenes. Chirality is lost during the reaction so enantiomeric substrates will give identical alkenes in both *syn*- and *anti*-eliminations. Like substitution reactions *syn*- and *anti*-eliminations are under *stereoelectronic control*; the nature of the reaction pathway determines the stereochemistry.

There are many examples of elimination reactions that are quite *unselective* and *non-specific*.

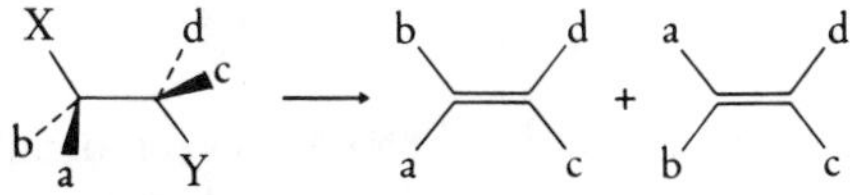

It is rare for these reactions to give exactly equal amounts of each diastereoisomer, but there are often significant amounts of each isomer. This can be inconvenient in synthesis if one diastereoisomer is required; but separation is usually tedious rather than difficult.

The reactions above are all assumed to be regiospecific as in each example given, only X and Y can eliminate. That is not always so; when Y is H there are often several possible elimination routes.

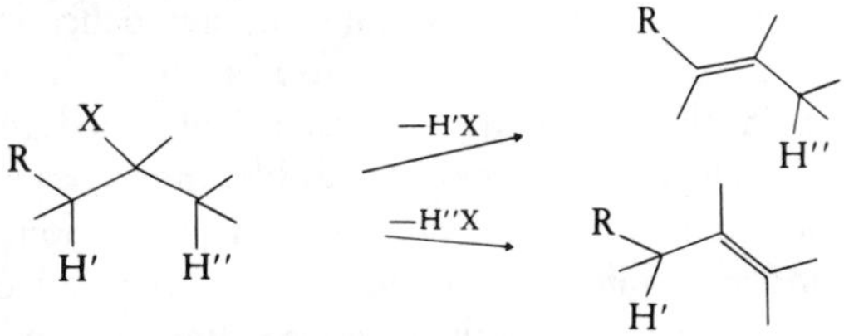

Such eliminations can be regiospecific, stereospecific, stereoselective or non-selective, depending on the structure of the substrate.

The mechanism of elimination reactions will now be examined according to the stereochemical relationship of X and Y leading to elimination. First, we shall concentrate on reactions with only one possible regiochemistry.

12.2 Stereospecific and stereoselective *anti*-eliminations

Most *anti*-eliminations are analogous to S_N2 substitutions as they are bimolecular, following a rate law,

rate = k[substrate] [base]
or rate = k[substrate] [nucleophile]

Such eliminations are called $E2$ for elimination, bimolecular. An example of a *base-induced anti-elimination* is,

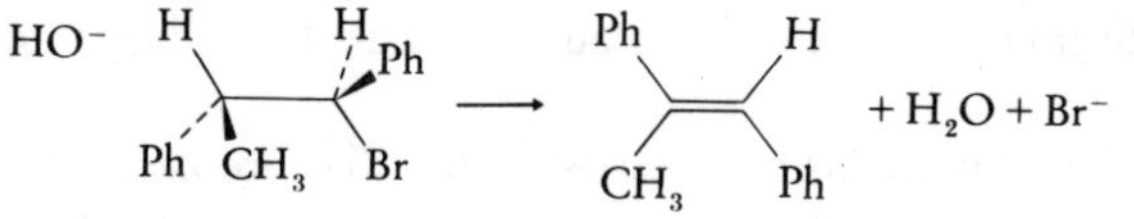

This reaction is induced, or assisted, or promoted by base but *not* catalysed by base as the base is consumed during the reaction.

A *nucleophile*-induced *anti*-elimination is usually associated with dehalogenation,

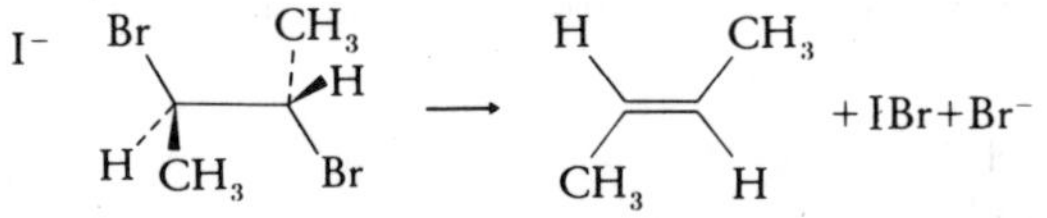

Stepwise eliminations are not ruled out by the rate laws but detailed kinetic studies have shown that anti-eliminations are concerted, passing through a single transition state. The mechanism can be illustrated.

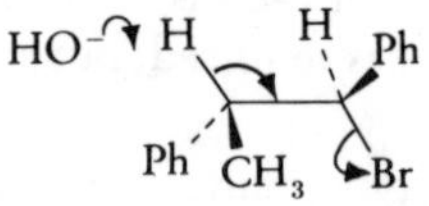

The questions of most stereochemical interest concern the *anti*-relationship of leaving groups in such eliminations.

Cyclohexyl derivatives can be used to reveal some stereochemical features of reactions. When a tertiary butyl group is used to lock the ring into a particular conformation the following results are obtained for tosylate eliminations. *cis*-4-*t*-Butylcyclohexyl tosylate readily undergoes $E2$ elimination with sodium ethoxide in ethanol. Under the same conditions the *trans*-isomer does not undergo bimolecular elimination but solvolysis, accompanied by some unimolecular elimination (see later). It can be seen by inspection of Figure 12.1, or models, that only the *cis*-isomer has the tosylate and a hydrogen atom in an antiperiplanar relationship. A similar experiment that makes the same point is the nucleophile-induced elimination of *cis* and *trans*-1,2-disubstituted cyclohexanes. These eliminations are shown schematically in Figure 12.2. The *trans*-1,2-disubstituted cyclohexane can only adopt one conformation in which C—X and C—Y are all coplanar.

In chair conformations, only the *trans*-diaxial arrangement satisfies the coplanarity condition. Any *trans*-1,2-disubstituted cyclohexane exists as an equilibrium between the *trans*-diaxial and the (usually more energetically favourable) *trans*-diequatorial conformer. It is the *trans*-diaxial conformation that leads to rapid elimination by an $E2$ mechanism. Models will demonstrate that a *trans*-1,2-disubstituted cyclohexane cannot be forced into a conformation in which C—X and C—Y are synperiplanar.

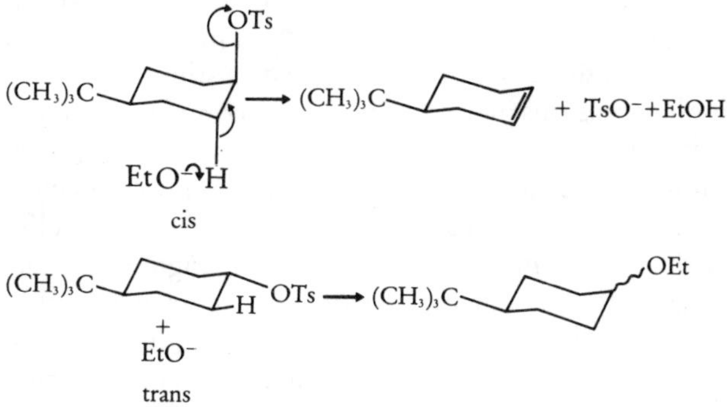

Figure 12.1 Reaction of *cis* and *trans*-4-*t*-butylcyclohexyl tosylate with ethoxide ion.

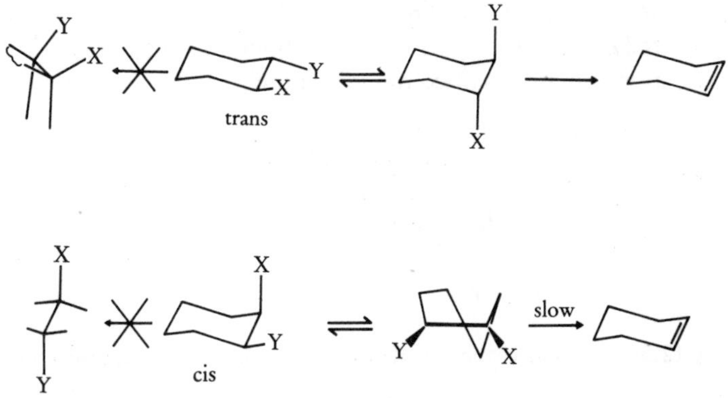

Figure 12.2 Nucleophile-induced E2 elimination of *cis* and *trans*-1,2-disubstituted cyclohexanes.

The $E2$ elimination of XY in the diastereoisomeric *cis*-1,2-disubstituted isomer is very slow; usually several powers of ten slower than the *trans* isomer. This is understandable if $E2$ elimination is preferentially *anti*. In the *cis*-isomer it is not possible to force the X and Y groups antiperiplanar, but they can be made synperiplanar in a twist-boat conformation. It appears that in the absence of a suitable antiperiplanar conformation, elimination can advance through the synperiplanar conformation; but only if there are no other lower energy pathways.

The rationalization of the detailed stereochemistry of elimination is not always obvious. There is no surprise that C—X and C—Y need to be coplanar for elimination. A π-bond, consisting of two parallel and sideways-overlapping p orbitals is formed during elimination. The lowest energy pathway for reaction will be the one where the maximum amount of π-bonding is developed during the reaction. This condition can only be met when C—X and C—Y are coplanar and the p-orbitals that form the new bond are already parallel. This applies equally to the *syn* and *anti* conformations. Explanation of the preference for *anti*-elimination is more difficult. As with S_N2 reactions there is a certain degree of justification for assuming that dipole–dipole repulsions play some part in determining that the entering base and departing group be as far removed as possible. The steric effect is probably influential in deciding the rapidity of *anti*-eliminations compared with *syn*-eliminations. It is also possible that there is more effective orbital overlap for the *anti*-elimination transition state. Figure 12.3 shows how the orbitals overlap at the transition state. As the geometry changes during elimination the p-orbitals take more part in π-bonding. Synchronous with this is the formation of the B—X bond and the loosening of the C—Y bond. In the *anti*-arrangement the overlap of the p-orbital with the C—Y bond may help to break that bond more rapidly than it could in a *syn* arrangement. Molecular orbital calculations also suggest that the overlap is better at half reaction for *anti*- than *syn*-elimination. The simplest calculations are based on a concerted elimination in which B—X bond making the C—Y bond breaking are approximately half completed at the transition state.

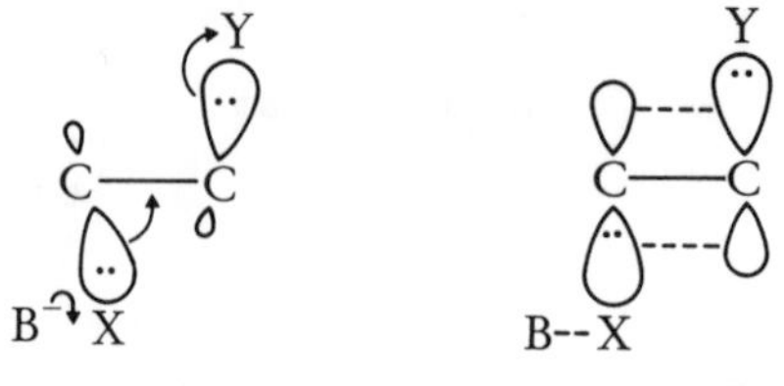

Figure 12.3   Qualitative picture of the orbital overlap in E2 transition states.

It is not necessary for bond-breaking and bond-making to be completely synchronous. In many $E2$ eliminations charge is developed in the transition state. Sometimes the bond to the proton is formed to a much greater extent than the bond to the other leaving group is weakened. In this case the transition state will bear a greater negative charge than the substrate. In the opposite case the loss of the leaving group, Y^-, is more advanced than bonding to the other leaving group. The transition state will therefore bear a greater positive charge than the substrate. Therefore without knowing the

162

mechanism in detail it is not possible to predict what effect electron-donating or releasing substitutents will have on the rate of an $E2$ elimination.

$E2$ eliminations are stereospecific and the following pairs of reactions are typical examples,

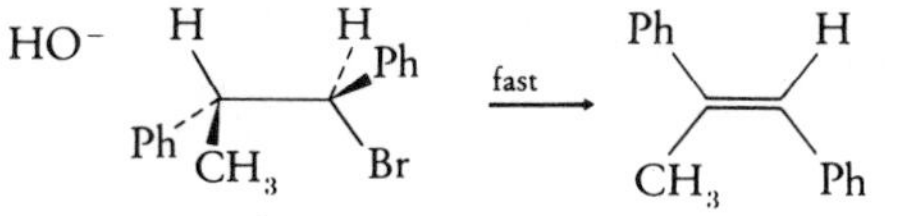

(12.1)

threo

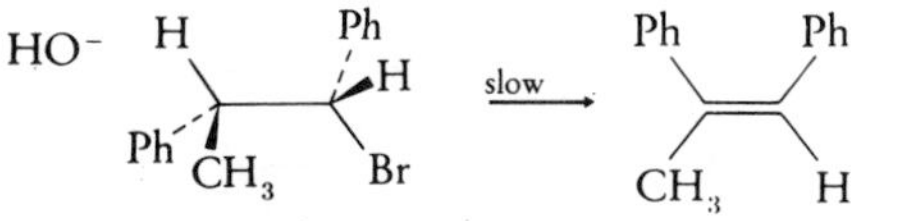

(12.2)

erythro

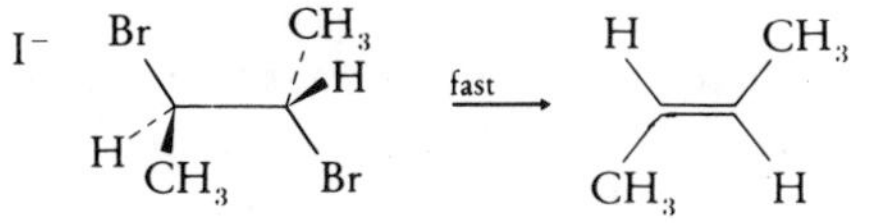

(12.3)

meso

(12.4)

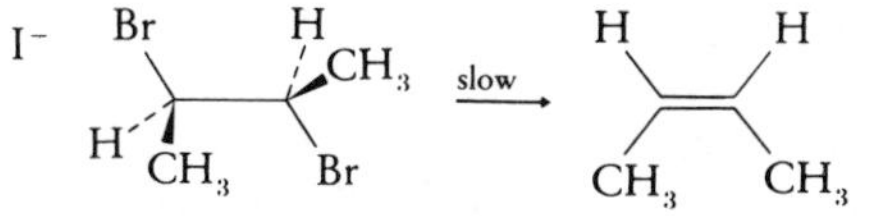

d, ℓ pair

In both of these examples there is a moderate rate difference for elimination from the diastereoisomers. The current explanation for the relative rates, 1 > 2 and 3 > 4 is based on the difference in steric crowding in the transition states leading to elimination. If we consider the eliminations in equations (1) and (2), the argument can be advanced. The phenyl groups are the bulky groups with a greater steric effect than the methyl group. The Newman projection shown in Figure 12.4 shows how the steric crowding in the transition state for the elimination of the *erythro* bromide is greater than in the ground state. The difference in energy between the ground states of the *threo* and *erythro* bromides is measurable but much smaller than the difference in energies between the respective elimination transition states. There is a greater activation energy for elimination from the *erythro* bromide, and a consequently lower rate. A similar argument can be presented for the faster elimination of the *meso* dibromide (equation 3) than the *d,ℓ* dibromides (equation 4).

Figure 12.4 *left erythro* bromide; *right* transition state geometry for elimination of HBr from the *erythro* bromide.

In the eliminations described so far there has only been one possible *anti*-elimination product for a given compound, so the reactions have been stereospecific. Under different circumstances *anti*-eliminations can be *stereoselective*

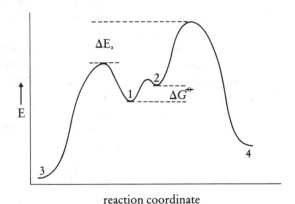

The reaction profile for this elimination is given in Figure 12.5.

The kinetic argument to account for the difference in the rates of elimination of diastereoisomers has been formalized, to account for product preference in *competing* reactions. This is the *Curtin–Hammett principle,* that states that for competing reactions the *product distribution is determined by the difference in activation energies for the two paths* and does not reflect the difference in energies between conformations, when this is small compared with the activation energies. This principle is used to explain why a conformer present in minor amounts can lead to the major product. In the example of Figure 12.5 the conformation present in the greatest amount leads to the lowest energy product by the most favourable route. In other words the product formed is both thermodynamically and kinetically controlled. There are however many other cases where the Curtin–Hammett principle is more useful!

Figure 12.5 Reaction coordinate diagram for the E2 elimination of HCl from PhCH$_2$CHClPh.

12.3 Stereospecific *syn*-eliminations

Syn-eliminations are generally unimolecular, by contrast to *anti*-eliminations. The reactions are accepted to take place through a concerted, cyclic mechanism. The reaction for *xanthate* elimination is,

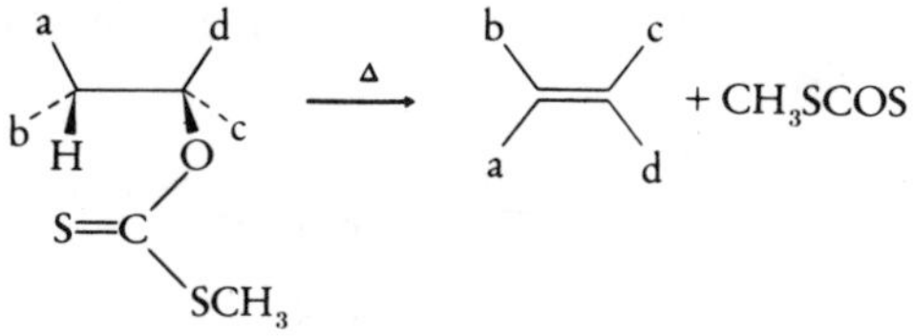

In addition to the xanthate, $SCH_3COS—$ group, the carboxylate,

$$RCOO^-,$$

and amine oxide $—\overset{\overset{\displaystyle O^-}{|}}{N}R_2$ groups can also be used to induce *syn* eliminations. The *syn*-stereochemistry of elimination can be demonstrated by using cyclohexane derivatives,

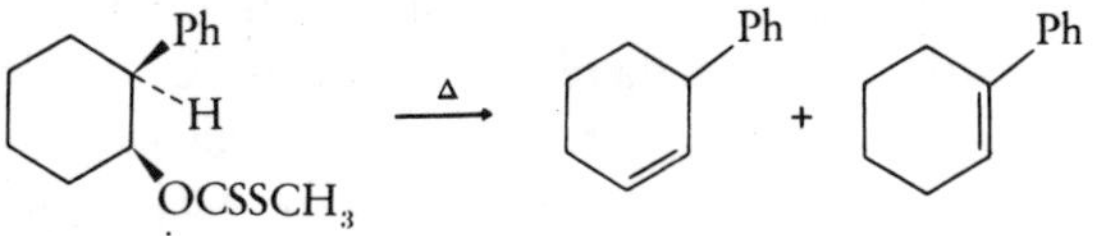

The thermodynamically more stable product is the conjugated, 1-phenylcyclohexene, **6**. The xanthate group and the hydrogen adjacent to the phenyl group can have an antiperiplanar (antidiaxial) relationship, but not a synperiplanar relationship. It is easy for the xanthate and one of the neighbouring methylene protons to attain coplanarity with the two directly bonded carbon atoms. This leads to the thermodynamically less-favoured product, 3-phenylcyclohexene, **5**, that is formed in large excess. The reaction is under stereoelectronic control and **5** is formed by kinetic control. The fact that up to 4% of **6** is produced is typical of *syn*-eliminations which are not always exclusively stereospecific or stereoselective.

The *trans* isomer leads to a fair amount of **5**. In this case both **5** and **6** can be formed by *syn*-elimination and it may be a steric preference that is expressed.

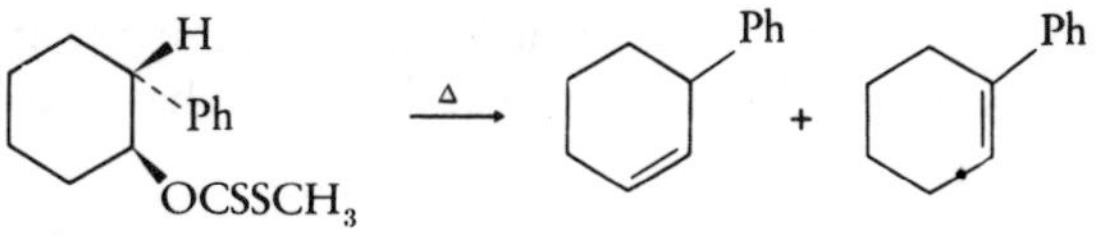

The regioselectivity of *syn*-eliminations is covered in Section 12.5.

12.4 Non-stereoselective eliminations

There are a group of unimolecular, non-stereospecific and non-stereo-selective eliminations that are classed together as $E1$ eliminations. They are analogous and related to, S_N1 reactions and they proceed by a carbonium ion intermediate. The carbonium ion is common to S_N1 and $E1$ reactions, with subsequent attack by nucleophile in S_N1, and loss of a proton to produce elimination in an $E1$ reaction.

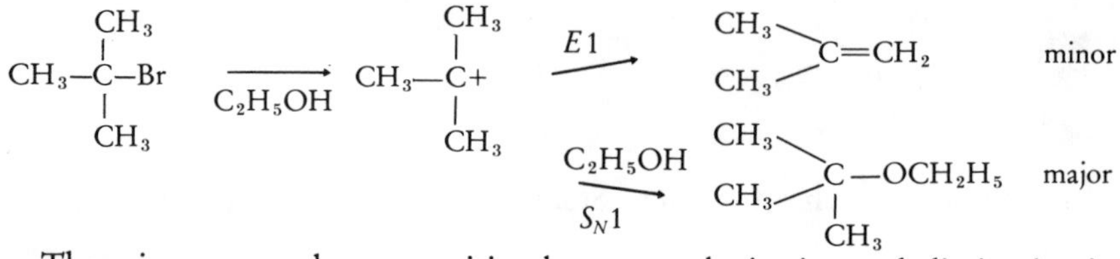

There is more on the competition between substitution and elimination in Section 12.6.

The rate law for $E1$ eliminations is,

Rate $= k[\text{substrate}]$

Elimination follows S_N1 solvolysis in being more effective in polar solvents and with tertiary substrates. When there is a choice of groups to eliminate, $E1$ eliminations are not stereospecific, stereoselective or regioselective.

There are examples of $E1$ eliminations where only one pair of ligands can eliminate, but diastereoisomers can be formed. In these examples diastereoisomeric substrates give the same product mixture, as the carbonium ion formed is free to rotate before elimination of a proton.

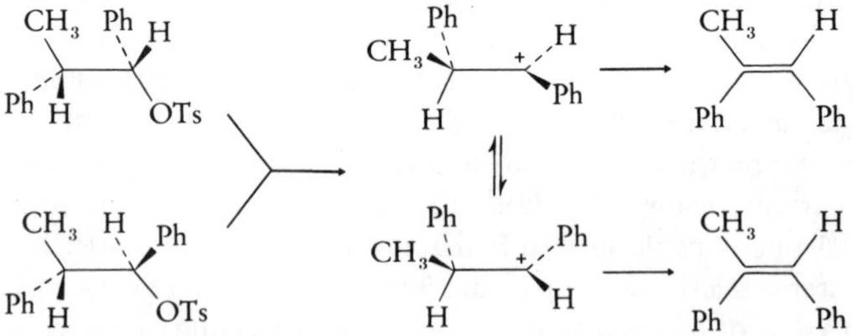

The more stable alkene is formed preferentially, but the E to Z ratio is usually determined by kinetic control and does not necessarily reflect the free energy difference between diastereoisomers.

A complication that can arise in $E1$ elimination reaction, and in S_N1 solvolyses, is that the intermediate carbonium ion can *rearrange*, so that the products are constitutional isomers of those that might have been expected,

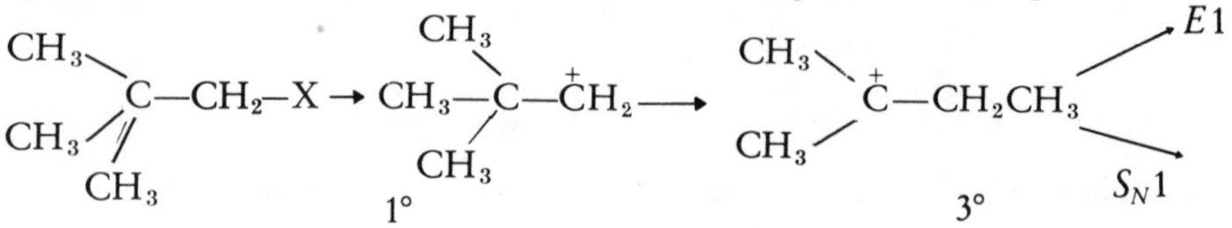

Tertiary carbonium ions are formed most readily by rearrangement, as the order of carbonium ion stability is, $3° > 2° > 1° > CH_3^+$.

12.5 Regioselectivity-Hofmann and Saytzeff elimination

In elimination reactions of the type,

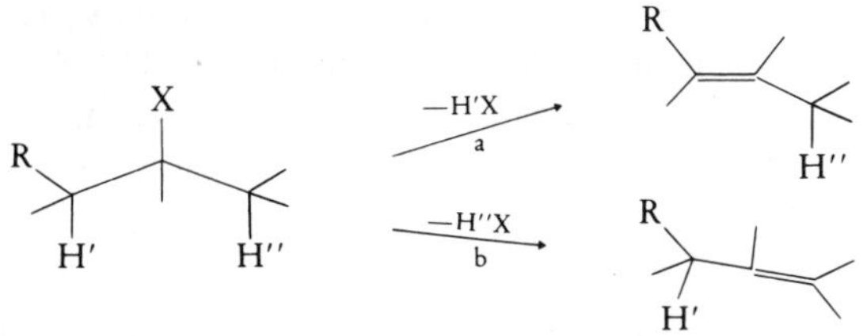

where two (or perhaps more) pathways are possible, two types of preference can be observed. If the *most substituted alkene* is formed preferentially the *Saytzeff rule* is said to be followed. If the least *substituted alkene* is formed preferentially the *Hofmann* rule is followed.

$E2$ eliminations from *neutral* substrates tend to follow the Saytzeff rule,

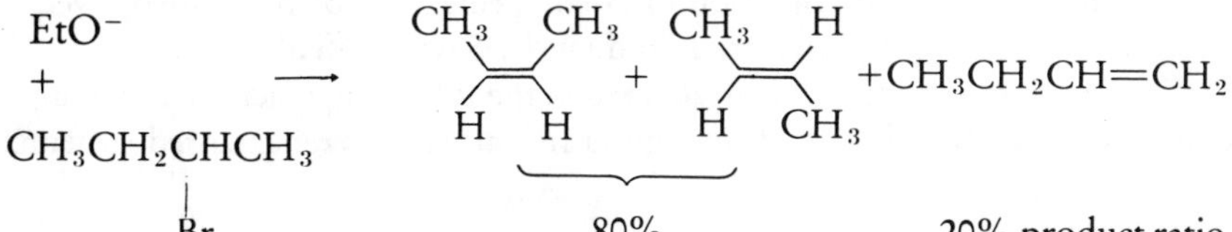

It is argued that in these examples the transition state has considerable double bond character and the product distribution therefore reflects the stability of the product alkenes.

$E1$ and *syn*-eliminations also usually give a majority of the Saytzeff product,

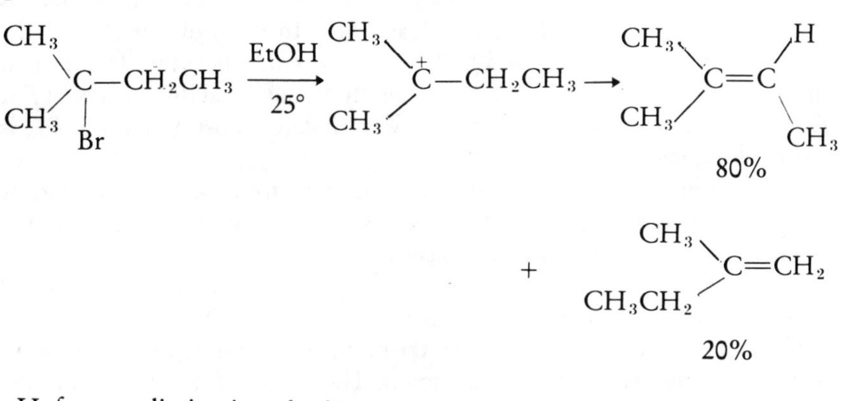

Hofmann elimination, leading to the least substituted alkene is found predominantly in $E2$ eliminations of charged substrates.

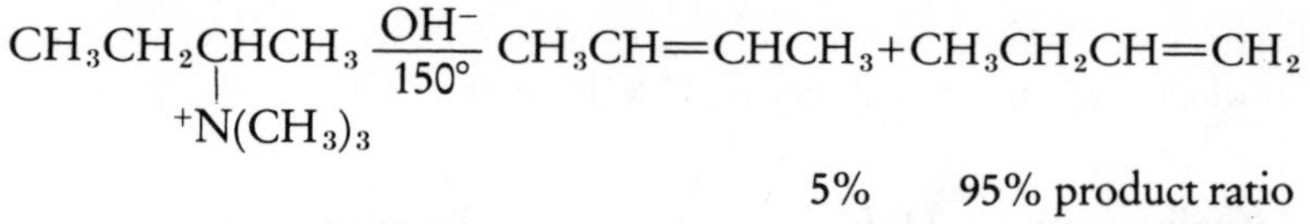

$$CH_3CH_2\underset{\underset{\displaystyle {}^+N(CH_3)_3}{|}}{C}HCH_3 \xrightarrow[150°]{OH^-} CH_3CH=CHCH_3 + CH_3CH_2CH=CH_2$$

5%   95% product ratio

Kinetic studies have shown that the transition state in the $E2$ elimination of quaternary ammonium salts has considerable carbanion character. This is consistent with the observed regiochemistry, as the most acid protons are those related to the most stable carbanions, where 1° > 2° > 3°. In consequence the least substituted alkene is formed. There may also be a steric element in Hofmann eliminations, as the hydrogen atoms removed are in the most accessible, as in,

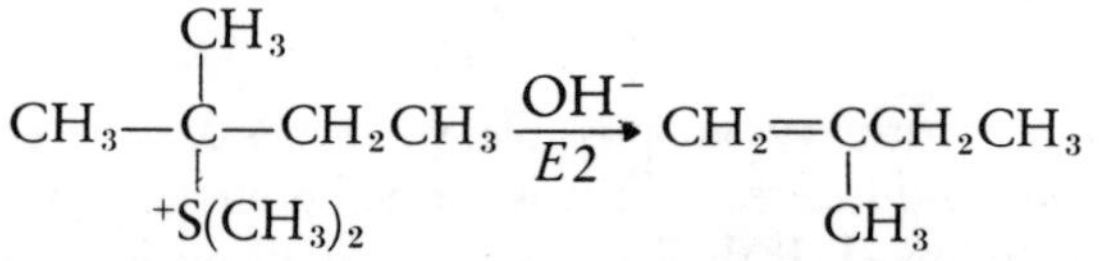

$$CH_3-\underset{\underset{\displaystyle {}^+S(CH_3)_2}{|}}{\overset{\overset{\displaystyle CH_3}{|}}{C}}-CH_2CH_3 \xrightarrow[E2]{OH^-} CH_2=\underset{\underset{\displaystyle CH_3}{|}}{C}CH_2CH_3$$

Eliminations expected to follow either Hofmann or Saytzeff orientation will not do so if there are severe stereochemical or electronic constraints. Bridgehead alkenes, in violation of Bredt's rule, cannot be formed, even if they are the apparently preferred products by Saytzeff's rule.

Similarly, Hofmann's rule is violated in the following reaction as conjugation lowers the transition state energy sufficiently for styrene to predominate,

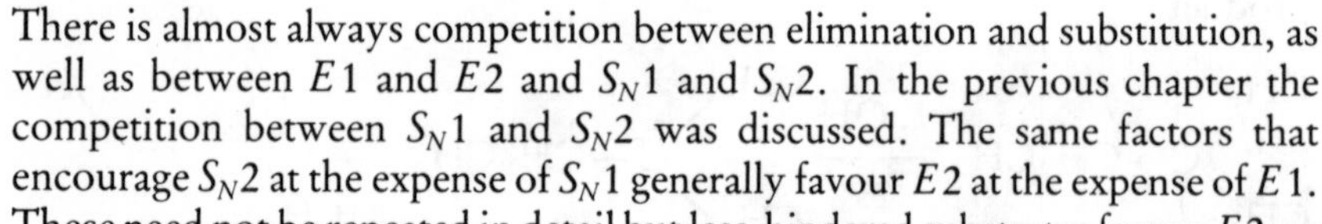

$$C_6H_5CH_2CH_2-\underset{\underset{\displaystyle CH_3}{|}}{\overset{\overset{\displaystyle CH_3}{|}}{{}^+N}}-CH_2CH_3 \xrightarrow[150°]{OH^-} C_6H_5CH=CH_2 + CH_2=CH_2$$

96%   4% product ratio

12.6 Elimination versus substitution

There is almost always competition between elimination and substitution, as well as between $E1$ and $E2$ and S_N1 and S_N2. In the previous chapter the competition between S_N1 and S_N2 was discussed. The same factors that encourage S_N2 at the expense of S_N1 generally favour $E2$ at the expense of $E1$. These need not be repeated in detail but less-hindered substrates favour $E2$, as do non-ionizing solvents.

The reaction conditions can therefore be made to favour bimolecular, or unimolecular reaction. The problem now is how to change the substitution to elimination ratio for each of these classes.

Bimolecular reactions will be tackled first, with discussion limited to $E2$ eliminations of HX competing with S_N2 substitutions of X by Y.

The main mechanistic point is that the group Y^- is acting as a *nucleophile* in substitution but as a *base* in elimination. The nature of Y^- is probably the biggest determinant in deciding the substitution to elimination ratio. Non-basic nucleophiles, such as I^-, Cl^-, and CN^- will strongly favour S_N2

reactions. Elimination is promoted by non-nucleophilic, strong bases. Steric hindrance in Y^- reduces the nucleophilicity while having less effect on the basicity. The effectiveness in promoting elimination is,

$$(CH_3)_3CO^- > (CH_3)_2CHO^- > CH_3CH_2O^-$$

The substrate structure, although not always able to be varied, if a given product is required, is important for bimolecular reactions.

$$C_2H_5O^- + C_2H_5Br \rightarrow C_2H_5OC_2H_5 + CH_2{=}CH_2$$
$$\quad\quad\quad\quad\quad\quad\quad\quad 90\% \quad\quad\quad 10\%$$

$$C_2H_5O^- + (CH_3)_2CHBr \rightarrow (CH_3)_2CHOC_2H_5 + CH_3CH{=}CH_2$$
$$\quad\quad\quad\quad\quad\quad\quad\quad\quad\quad 20\% \quad\quad\quad\quad 80\%$$

$$C_2H_5O^- + (CH_3)_3CBr \rightarrow {-}(CH_3)_2C{=}CH_2$$
$$\quad\quad\quad\quad\quad\quad\quad\quad \sim 100\%$$

Tertiary substrates do not undergo S_N2 reactions so $E2$ eliminations are almost exclusive with strong bases (although some S_N1 solvolysis may occur concurrently).

Charged leaving groups also seem to favour elimination,

$$C_2H_5O^- + C_2H_5Br \rightarrow CH_2{=}CH_2 + C_2H_5OC_2H_5$$
$$\quad\quad\quad\quad\quad\quad\quad 1\% \quad\quad\quad 99\%$$

$$C_2H_5O^- + C_2H_5\overset{+}{N}(CH_3)_2 \rightarrow CH_2{=}CH_2 + C_2H_5OC_2H_5$$
$$\quad\quad\quad\quad\quad\quad\quad\quad\quad 70\% \quad\quad\quad 30\%$$

The activation energy is higher for elimination than for substitution; higher temperatures therefore promote elimination.

The competition between substitution and elimination under unimolecular conditions is less easy to control, as S_N1 and $E1$ reactions have a common carbonium ion intermediate.

In moderately polar solvents the ratio of substitution to elimination is relatively independent of the leaving group,

$$(CH_3)_3CBr \xrightarrow[20\% \ H_2O]{80\% \ C_2H_5OH} (CH_3)_2C{=}CH_2 +$$
$$\quad\quad\quad\quad\quad\quad\quad\quad\quad\quad\quad\quad 36\%$$

$$(CH_3)_3COH + (CH_3)_3COC_2H_5$$
$$\quad\quad 64\%$$

$$(CH_3)_3C\overset{+}{S}(CH_3)_2 \xrightarrow[20\% \ H_2O]{80\% \ C_2H_5OH} (CH_3)_2C{=}CH_2 +$$
$$\quad\quad\quad\quad\quad\quad\quad\quad\quad\quad\quad\quad 36\%$$

$$(CH_3)_3COH + (CH_3)_3COC_2H_5$$
$$\quad\quad 64\%$$

Substitutions most frequently predominate under unimolecular conditions.

The conclusion is that if control of substitution to elimination is required, bimolecular conditions make that control more easy.

12.7 Summary of Sections 12.1–12.6

1. Stereospecific eliminations can be *anti* or *syn*

$$\text{anti,} \qquad \overset{X}{\underset{a}{\diagup}}\!\!\overset{d}{\underset{b}{\diagdown}}\!\!\overset{c}{\underset{Y}{}} \longrightarrow XY + \overset{a}{\underset{b}{\diagup}}\!\!=\!\!\overset{d}{\underset{c}{\diagdown}}$$

$$\text{syn,} \qquad \overset{X}{\underset{a}{\diagup}}\!\!\overset{Y}{\underset{b\ \ d}{}}\!\!\overset{}{\underset{c}{}} \longrightarrow XY + \overset{a}{\underset{b}{\diagup}}\!\!=\!\!\overset{c}{\underset{d}{\diagdown}}$$

2. Most *anti*-eliminations are bimolecular $E2$ eliminations with rate = k[substrate] [base] for HX elimination, and rate = k[substrate] [nucleophile] for XY elimination. $E2$ eliminations are concerted and are induced by base or nucleophile.
3. *Anti*-elimination is favoured on steric and electronic grounds. The orbital overlap at the transition state for base promoted elimination is better for *anti*-elimination than *syn*.
4. Differences in steric crowding in the transition states accounts for the difference rates of elimination of diastereoisomers. The least hindered eliminates more rapidly and crowding is greater in the transition state than in the ground states.
5. When different conformations can lead to different products, the product with the lowest activation energy for formation will dominate, as long as the activation energies are large compared to the difference in energy of the conformations. This is the Curtin Hammett principle.
6. *Syn*-eliminations are usually unimolecular and concerted, requiring a synperiplanar arrangement.
7. $E1$ eliminations, analogous to S_N1 reactions, are non-specific. The intermediate carbonium ion can rotate before elimination and a mixture of diastereoisomers is formed.
8. When eliminations are not necessarily regioselective one of two rules is usually followed. Saytzeff rule states that the *most* substituted alkene will be formed. The Hofmann rule states that the *least* substituted alkene will be formed.
9. The Saytzeff rule is followed by: $E2$ eliminations of neutral substrates; *syn*-eliminations and most $E1$ eliminations. The transition state for $E2$ eliminations is alkene-like and the final product reflects the greater stability of substituted alkenes.
10. Hofmann elimination occurs in $E2$ elimination of charged substrates.

The transition state has carbanion character and the alkene formed mirrors the intermediate carbanion stability.

11. Reaction conditions can be adjusted to favour S_N1 and $E1$ relative to S_N2 and $E2$, when competition is possible.
    $E1$ and $E2$ are favoured by the same factors as S_N1 and S_N2 as the transition states are similar.

12. In $E2$ vs S_N2, strong *bases*, tertiary substrates and high temperatures all favour $E2$.

13. In $E1$ vs S_N1 it is difficult to change the ratio as the two sets of products are formed from a common intermediate, formed in the rate-determining step.

12.8 Addition reactions

In principle addition reactions are the reverse of elimination reactions.

$$>C{=}C< + HX \rightleftharpoons >CHC(X)<$$

$$>C{=}C< + XY \rightleftharpoons >CXC(Y)<$$

In practice the conditions will be different as elimination reactions are often carried out in the presence of base to remove HX and prevent reversibility.

The additions examined here are *regiospecific*, *stereospecific* and either *anti* or *syn*.

12.8.1 Anti *additions*
Addition of halogens to alkenes are examples of *electrophilic, anti additions*.

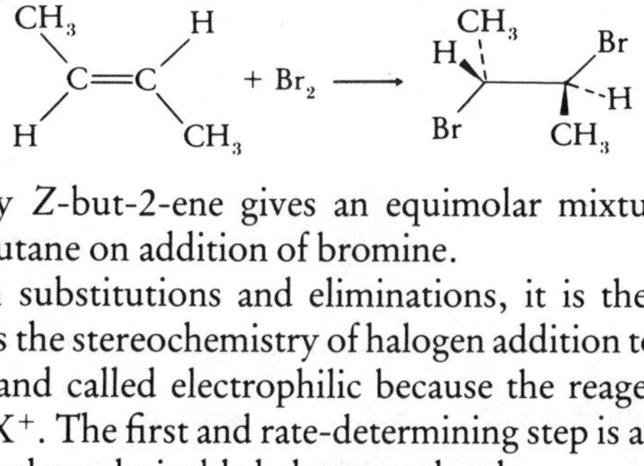

Similarly Z-but-2-ene gives an equimolar mixture of R,R and S,S-2,3-dibromobutane on addition of bromine.

As with substitutions and eliminations, it is the reaction character that determines the stereochemistry of halogen addition to alkenes. The reaction is stepwise, and called electrophilic because the reagent halogen behaves as a source of X^+. The first and rate-determining step is attack by the nucleophilic π-bond on the polarizable halogen molecule.

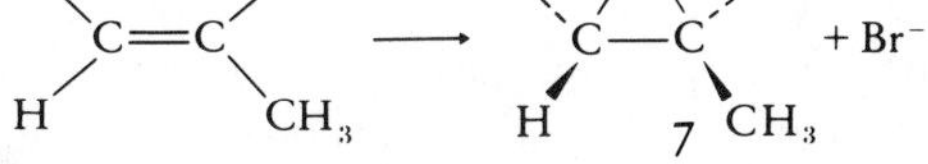

The ion 7 is called a *halonium ion* (*bromonium ion* in this case) in which the halogen atom is coordinated to *both* alkene carbon atoms. The configuration of the starting alkene is retained in this step. The attack can equally be at either face as there is an element of symmetry present in the reaction character. The existence of the halonium ion has been confirmed by nmr spectroscopy at low temperature.

The second step is the attack of halide ion, from the back to give overall *anti*-addition.

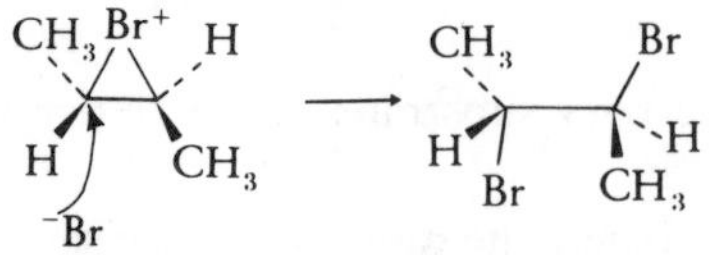

The attack from the rear is steric in origin, and leads to the most favoured conformation with bromine atoms antiperiplanar. In the example above bromide is shown attacking one particular carbon atom, but the result is the same whichever atom is attacked. There can be no preferred regiochemistry when adding a symmetrical reagent.

The stepwise addition can be confirmed by trapping the halonium ion intermediate with another nucleophile,

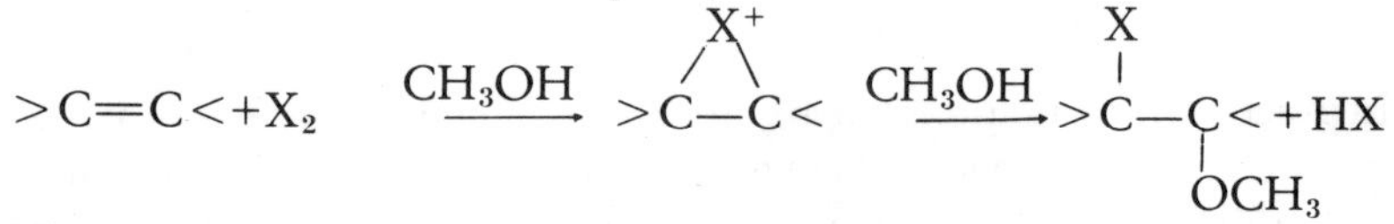

$$>C{=}C< + X_2 \xrightarrow{CH_3OH} >\overset{\overset{\displaystyle X^+}{\diagup\diagdown}}{C{-}C}< \xrightarrow{CH_3OH} >\overset{\overset{\displaystyle X}{|}}{C}{-}\underset{\underset{\displaystyle OCH_3}{|}}{C}< + HX$$

The preference for *anti*-addition is demonstrated by the addition of bromine to a 3-substituted cyclohexane, as shown in Figure 12.6.

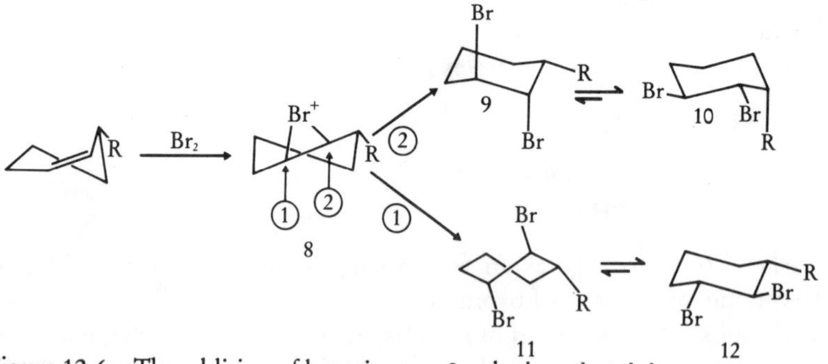

Figure 12.6　The addition of bromine to a 3-substituted cyclohexane.

The first step is the addition to give the bromonium ion. Owing to the chirality of the cyclohexene conformers there is discrimination of faces of the alkene (see Chapter 14). Bromine adds to the least hindered face preferentially, to give structure **8** which represents only one enantiomeric pair of the diastereoisomeric pairs of ions. The stereochemical consequences of attack at each carbon atom are different. Attack at carbon-2 is a relatively low energy process giving a chair cyclohexane, **9**, with diaxial bromine atoms. Attack at carbon-1 with overall *anti*-addition cannot lead to a chair conformation. A twist boat conformation, **11**, is formed in a relatively high energy process. Careful model manipulation can show this very effectively.

172

Both of the initially formed products, **9** and **11** are thermodynamically unstable with respect to other conformations. The diaxial bromide **9**, is in equilibrium with the preferred conformer **10**, and the twist boat **11** is of higher energy than the conformer **12**, with all groups equatorial. The thermodynamically preferred diastereoisomer is **12**, but the product of the reaction is **10**. This experiment shows both that *anti*-addition is preferred, and that kinetic control is in operation, as the lowest energy transition state leads to **9**.

The addition of HX to alkenes, where X is usually a halogen or OH, is largely *anti*, but the reaction is not so stereospecific as halogen addition. The proposed mechanism is similar to that for halogen addition. Instead of a bromonium ion a protonium ion is often proposed for the *initial* species.

$$>C\overset{H}{\underset{+}{\diagup}}C< \quad \text{or} \quad >C\overset{H^+}{=\!\!\!|\!=}C<$$

The *regiochemistry* of HX addition is of special interest synthetically, and can usually be predicted for electrophilic additions. The favoured product is the one resulting from the *most stable* carbonium ion. Carbonium ion stability follows the order,

$$3° > 2° > 1° > CH_3{}^+$$

$$CH_3CH\!=\!CH_2 + HCl \rightarrow CH_3\overset{\underset{|}{Cl}}{C}HCH_3$$

$$(CH_3)_2C\!=\!CHCH_3 + HI \rightarrow (CH_3)_2CICH_2CH_3$$

These reactions proceed through $CH_3\overset{+}{C}HCH_3$ and $(CH_3)_2\overset{+}{C}CH_2CH_3$ respectively. This regiochemistry is usually summarized as *Markownikov addition*. The original Markownikov rule stated that 'the hydrogen atom of HX adds to the carbon atom bearing the greater number of hydrogens'. For addition of H-halogen and H—OH this always results in the most stable carbonium ion intermediate. Care must be taken in using the Markownikov rule as the preferred *mechanistic* interpretation can suggest opposite results to the original formulation. This is illustrated in the next Section.

12.8.2 Syn *additions*

There are several kinds of *syn* additions but we shall only describe two. The first is *hydroboration*, which is a reaction of outstanding synthetic value, and won a Nobel prize for H.C. Brown, to whom the credit for its discovery and exploitation is due. The basic reaction is,

$$>C\!=\!C< + >B\!-\!H \rightarrow >\overset{\overset{\displaystyle >B}{\underset{\displaystyle |}{}}}{C}\!-\!\overset{\overset{\displaystyle H}{\underset{\displaystyle |}{}}}{C}<$$

The borane this formed can then be transformed into a series of different

products. One example is,

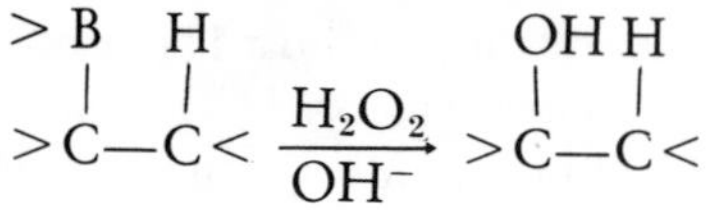

$$>\underset{\underset{>C-C<}{|\ \ \ |}}{B\ \ \ H} \xrightarrow[OH^-]{H_2O_2} >\underset{\underset{>C-C<}{|\ \ \ |}}{OH\ H}$$

The addition is *syn* but the mechanism is still not definitively described, but could be either concerted,

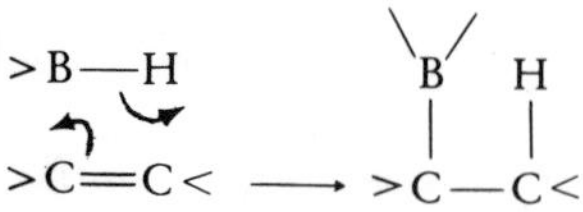

$$>\underset{>C=C<}{B-H} \longrightarrow >\underset{\underset{C-C<}{|\ \ |}}{B\ \ H}$$

or be an electrophilic addition to give an ion pair that collapses rapidly, in the same way as in the S_Ni reaction,

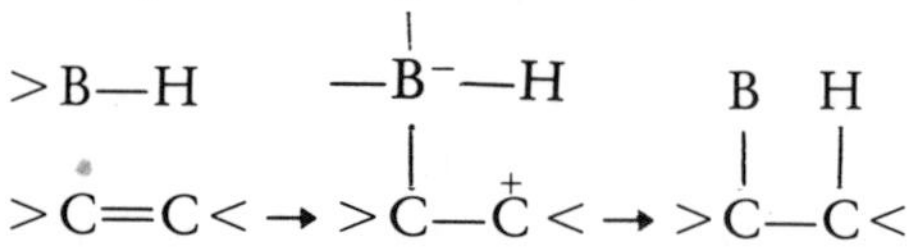

$$>B-H \quad -\overset{|}{B}^- -H \quad B\ \ H$$
$$>C=C< \rightarrow >\overset{}{C}-\overset{+}{C}< \rightarrow >\underset{}{C}-\underset{}{C}<$$

This stepwise reaction is useful for rationalizing the regiochemistry,

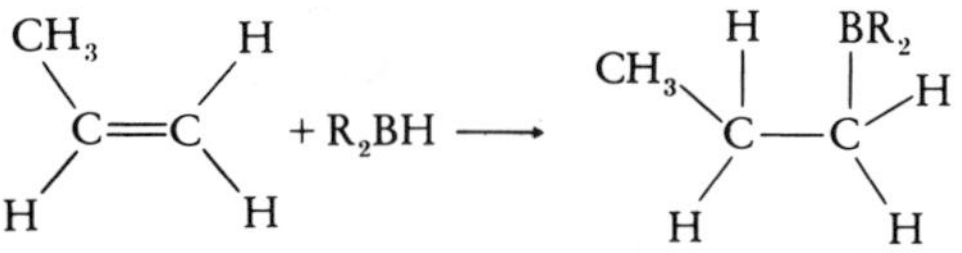

$$\underset{H}{\overset{CH_3}{>}}C=C\underset{H}{\overset{H}{<}} + R_2BH \longrightarrow \underset{H}{\overset{CH_3}{>}}\underset{}{C}-\underset{H}{\overset{BR_2}{C}}$$

According to the original Markownikov rule this addition is unexpected and is called anti-Markownikov addition, as the hydrogen adds to the *most* substituted carbon atom. The product formed is, however, the one expected from a mechanistic interpretation of the Markownikov rule. The boron atom is acting as the electrophilic centre and the hydrogen atom is behaving as a hydride, H^-, ion. The most stable carbonium ion, or carbonium-ion-like transition state results in the regiochemistry shown.

Hydroboration can be used effectively to enable an overall anti-Markownikov addition to a double bond,

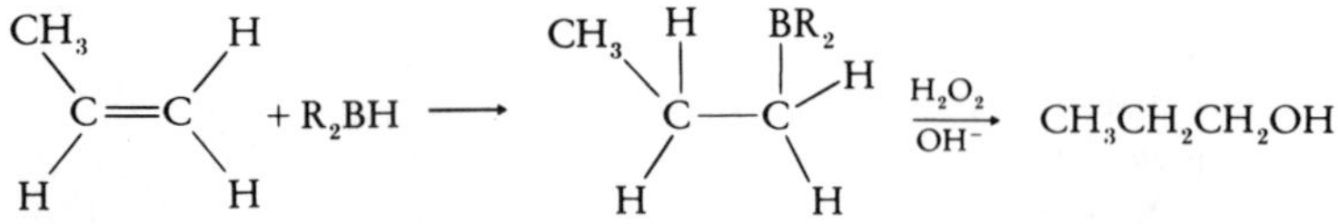

$$\underset{H}{\overset{CH_3}{>}}C=C\underset{H}{\overset{H}{<}} + R_2BH \longrightarrow \underset{H}{\overset{CH_3}{>}}\underset{H}{\overset{H\ \ BR_2}{C}-C} \xrightarrow[OH^-]{H_2O_2} CH_3CH_2CH_2OH$$

compared with,

$$\underset{H}{\overset{CH_3}{>}}C=C\underset{H}{\overset{H}{<}} + H_2O \xrightarrow{H^+} CH_3CH(OH)CH_3$$

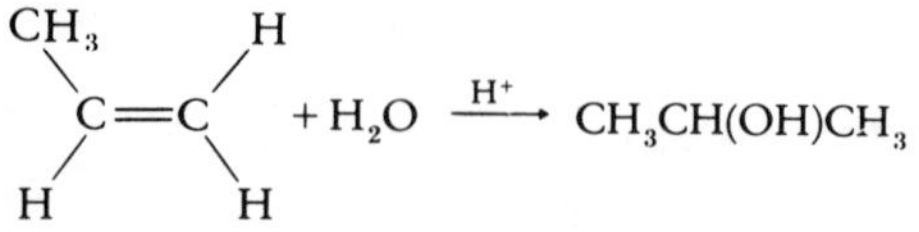

Syn addition of hydrogen to alkenes is observed in transition metal catalysed reductions with hydrogen. Platinum or palladium catalyse such reductions,

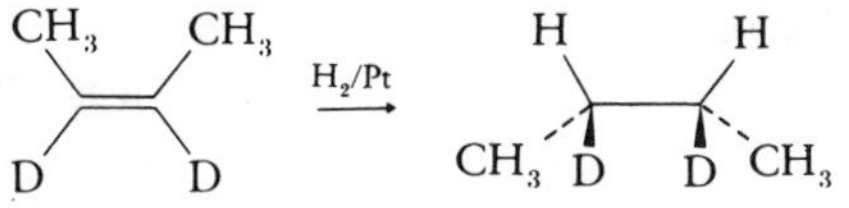

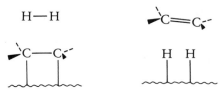

meso

The mechanism is complex as this is a *heterogeneous* process requiring coordination of the alkene, or hydrogen to the metal surface, as shown in Figure 12.7. The reactants are therefore orientated so that *syn* addition is inevitable.

Figure 12.7   Coordination of alkenes and hydrogen to metal surfaces results in catalysis of *syn* additions.

Transition metal complexes can also catalyse *syn* addition of hydrogen to alkenes.

12.8.3 Summary of Section 12.8

1. Addition reactions are usually regio and stereo selective and can be *syn* or *anti*.
2. Addition of halogen is *anti*, and two-step with intermediate halonium ion formation. The halonium ion is formed with retention of configuration at the alkene carbon atoms, 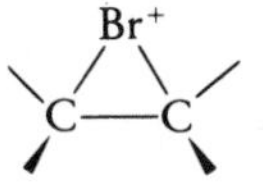 . Attack from the back by X⁻ (with inversion) is the second step to give the *anti*-dihalide.
3. Addition of HX is also *anti*, but not so stereospecific as halogen addition. Intermediate protonium ion formation is often postulated.
4. Electrophic addition of HX is regiospecific following Markownikov's rule that the hydrogen atom in HX adds to the carbon atom bearing the greater number of hydrogens. This is better expressed in a mechanistic sense by saying that the product from the most stable carbonium ion will be formed.
5. Hydroboration is addition of >B—H and is stereospecifically *syn* and the boron atom adds to the least-substituted carbon atom, consistent with intermediate ion-pair formation.
6. Transition metals catalyse *syn* additions of hydrogen to alkenes through a heterogeneous process.

Problems and Exercises

1. What are the most probable products of elimination from

     (a) at 200° where X = OOCCH₃
     (b) with conc. H_2SO_4 and X = OH
     (c) with EtO⁻ and X = Cl?

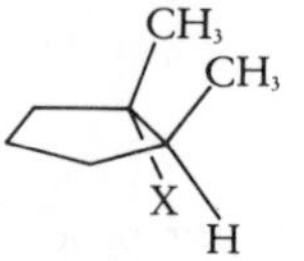

Write mechanisms for each elimination and rationalize your suggested product distribution.

2. What are the products of $E2$ elimination of HCl from **13** and **14**?

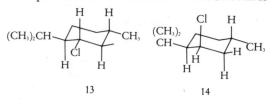

              13   **14**

Which product(s) predominate and why?

3. Which mechanism, S_N1, S_N2, $E1$ or $E2$ is likely to predominate for,
   - (a) $(CH_3)_2CHCHBrCH_3 + CN^- \longrightarrow$?
   - (b) $(CH_3)_2CHCHBrCH_3 + EtO^- \longrightarrow$?
   - (c) $(CH_3)_2CHCHBrCH_3 + Bu^tO^- \longrightarrow$?
   - (d) $(CH_3)_2CHCH\overset{+}{N}(CH_3)_3CH_3 + OH^- \xrightarrow{200°}$?

Write a structure for the major product.

4. Suggest elimination reactions to synthesize,

  (a)

         (b)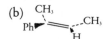

  (c)

5. Draw stereodiagrams of the major product(s) of the following additions.

  (a)

    (c) 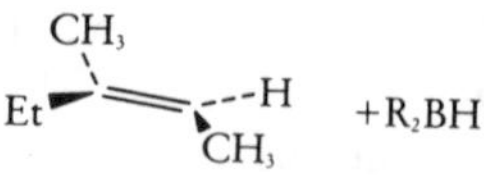 CH₃ ... Et =< ...H / CH₃   + R_2BH

  (b) + HBr, D

    (d)   + Br₂

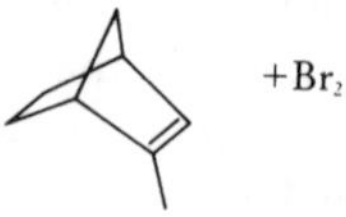

CHAPTER 13

Pericyclic reactions

13.1 Introduction

There are a large number of diverse additions, eliminations and rearrangements that are conveniently classed together as *pericyclic reactions*.

The basis of the classification is mechanistic and each pericyclic reaction satisfies four criteria. *They are concerted;* no radical or ionic intermediates can be detected. *The rates of pericyclic reactions are practically unaffected by either catalysts or changes in solvent. The reagents used are neither nucleophilic nor electrophilic.* The final criterion is the one that defines pericyclic reactions. At least two bonds are made and broken *in a single concerted step with a cyclic transition state.* These four mechanistic features set pericyclic reactions apart from the reactions studied in the previous two chapters.

We shall examine three classes of pericyclic reactions.

1. Electrocyclic ring openings and closures. An electrocyclic ring closure is one in which a σ-bond is developed between two termini of a conjugated π-system. The product therefore has one fewer π-bond than the reactant. The opposite of this process is an electrocyclic ring opening.

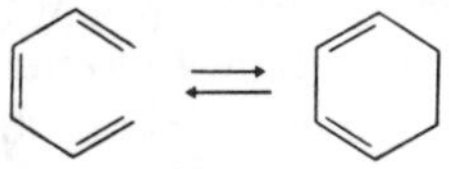

2. Cycloaddition reactions are reactions in which two σ-bonds are formed between the termini of two isolated π-systems.

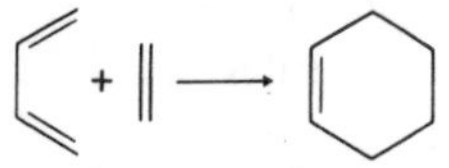

Cycloreversions are the opposite of cycloadditions.

3. Sigmatropic rearrangements involve the migration of a group, or atom, across a π-electron framework.

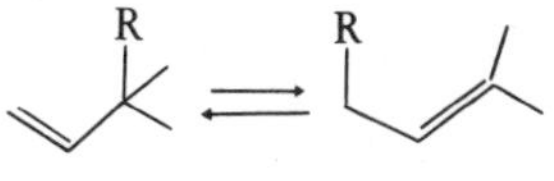

177

Reactions of these, and other classes, have been known for many years but until the mid 1960's proved to be difficult to describe mechanistically. At one time some chemists were so exasperated that they described them as 'no mechanism' reactions. Many pericyclic reactions are stereospecific and that provided strong stimulus for their study. In a brilliant series of papers R.B. Woodward (Nobel Prize for chemistry 1965), together with R. Hoffmann (Nobel Prize for chemistry 1980) expounded the theory of the *conservation of orbital symmetry* which rationalized the course of pericyclic reactions. Their original treatment used molecular orbital theory to produce a set of selection rules, now called the Woodward–Hoffmann rules. Following their work a number of 'back of an envelope' treatments of pericyclic reactions have appeared. These all lead to the same rules, but are conceptual models, rather than rigorous treatments. Woodward and Hoffmann described the basis of their method as 'the introconvertible proposition that a chemical reaction will proceed the more readily, the more *bonding* may be maintained throughout the transformation'. The treatment that we shall adopt is based on the premise that in pericyclic reactions *aromatic transition states* maintain the most bonding. Using this approach we shall explain, among other things, why ethene does not dimerize thermally

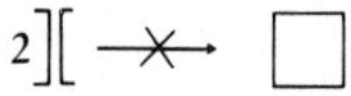

but butadiene adds readily to ethene,

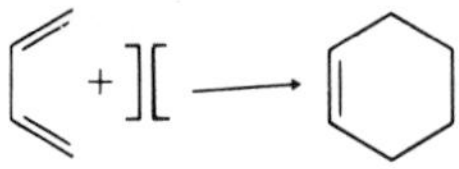

and why, in the following reaction, only one diastereoiomer is formed thermally

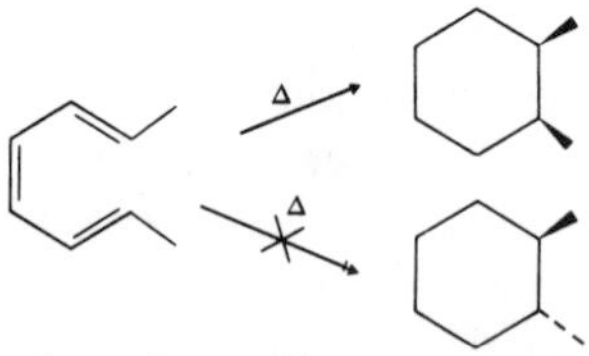

The Woodward–Hoffmann rules are applicable to these and many other reactions but it must be stressed that they refer only to *concerted reactions*. When the rules apparently fail for a particular reaction it is probable that it proceeds in a stepwise manner with radical or ionic intermediates. The possibility of non-concerted processes must always be borne in mind and it is a difficult matter experimentally to establish concertedness.

Before establishing the Woodward–Hoffmann rules, a brief revision and extension of the ideas of aromaticity is needed.

13.2 Aromaticity in the context of pericyclic reactions

13.2.1 Hückel aromaticity
Benzene is an especially stable conjugated compound, and calculations show

that it is much lower in energy than the hypothetical, undelocalised cyclo-hexatriene. Cyclobutadiene is much *destabilized* by interaction of the two π-bonds. These are two examples of the generalization that $4n+2$ electrons in a cyclic array are aromatic and $4n$ electrons are antiaromatic. A necessary condition of this description of aromaticity is that a number of p orbitals of the same type (p_x, p_y or p_z) overlap in a sideways manner.

A planar carbon skeleton is also implied.

Figure 13.1(a) illustrates the *basis set* of atomic orbitals used in the construction of the molecular orbitals for benzene. In this representation we have chosen to label the lobes on the same side of the molecular plane with the same sign. For the treatment we shall adopt this is unnecessary. The same *molecular orbitals* are obtained on combination of the atomic orbitals, no matter how the phase relationships in the basis set are displayed. The same basis set for benzene is shown in Figure 13.1(b), but the phases are randomly assigned. The wavy lines indicate *phase dislocations*, where a positive and negative lobe are adjacent. Every possible combination of phases in the benzene basis set gives rise to either zero or an even number of phase dislocations. Such a cyclic system with an even number of phase dislocations (zero, even) is called a *Hückel system*.

A Hückel system with $4n+2$ electrons is aromatic and with $4n$ electrons is antiaromatic.

(a)   (b)

Figure 13.1   (a) The benzene basis set of p orbitals arranged so that lobes of the same phase are adjacent; (b) the benzene basis set with randomly assigned orbital phases. The wavy lines indicate phase dislocations.

Throughout this chapter orbital phases are indicated by shading. Black lobes are of opposite phase to white lobes. Only the relative phase of orbitals is of interest.

13.2.2 Anti-Hückel aromaticity

Stable π-systems with an *odd* number of phase dislocations are unknown, and the only way that they can be visualized is through a twisted, or Möbius system. The simplest way to be convinced that a Möbius system has an odd number of phase dislocations is to make a paper model. Take a strip of paper, say 20 cm by 2 cm, draw a p orbital across the strip at each end, and put a random number of p orbitals between these. Now label the phases randomly. Twist *one* end of the paper through 180° and join the ends together. You now have a single surface Möbius strip. No matter how many p orbitals you have chosen, and however their phases are arranged, there will *always* be an odd number of phase dislocations. Such an array is called an *anti-Hückel* system. It can be shown that the condition for aromaticity in anti-Hückel systems is opposite to that for Hückel systems.

An anti-Hückel system with 4n electrons is aromatic and with 4n+2 electrons is anti-aromatic.

Except for very large rings, where aromatic stabilization is small anyway, the twist in the Möbius system reduces the π-electron overlap too much for the observation of stable molecules. Transition states may have twisted π-systems and their energy can be lowered by aromaticity.

Although the concepts described above apply stricty to planar molecules, the assumption made in our derivation of the Woodward–Hoffmann rules is that non-planar transition states can be 'aromatic' or anti-aromatic'. Transition states are examined for the number of electrons, and the number of phase dislocations. Each transition state can therefore be treated, to a first approximation, as being stabilized or destabilized by aromaticity. In general, for any pericyclic reaction, a choice of routes is possible and the one (or ones) leading to an 'aromatic' transition state will be favoured for thermal reactions.

13.3 Electrocyclic reactions

The simplest way to establish the Woodward–Hoffmann rules is to look at a few examples of pericyclic reactions and examine their transition states. We shall start with electrocyclic reactions.

The thermal electrocyclic ring opening of *cis*-3,4-dimethylcyclobutene to *E,Z*-hexa-2,4-diene is completely stereospecific,

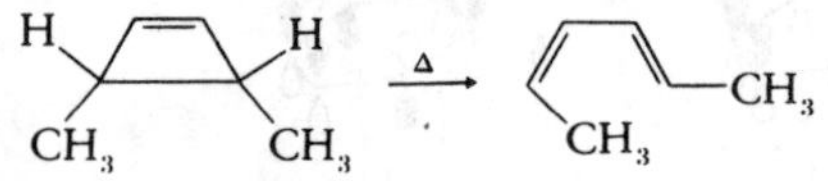

This reaction takes place by breaking the 3,4 C—C bond. There are two ways that this σ-bond could break, defined by the direction of rotation of the groups on each carbon atom, about the 2,3 and 1,4 C—C bonds. The two modes, which are general for electrocyclic reactions are called *conrotatory* and *disrotatory* and are illustrated in Figure 13.2. Conrotatory openings have both rotations in the same direction, shown here as clockwise (but equally likely to be anticlockwise, as each leads to the same product). Disrotatory openings have rotations in opposite directions. The disrotatory mode giving the *E,E* isomer for *cis*-3,4-dimethylcyclobutene thermolysis is shown in Figure 13.2.

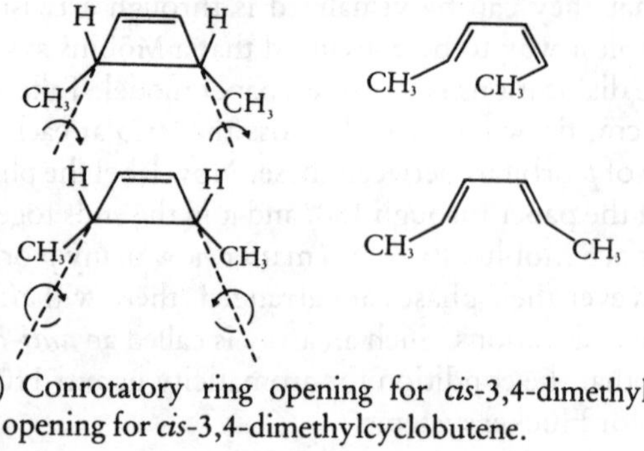

Figure 13.2 (a) Conrotatory ring opening for *cis*-3,4-dimethylcyclobutene. (b) Disrotatory ring opening for *cis*-3,4-dimethylcyclobutene.

180

There is another disrotatory mode leading to the Z,Z alkene that is not illustrated. The two disrotatory modes lead to diastereoisomers and are not therefore of equal energy or probability. This is mentioned again in the problems.

Now let us draw the basis sets for the transition states for conrotatory and disrotatory ring opening of cyclobutenes, and analyse them for aromaticity. Figure 13.3(a) illustrates the basis set for the starting cyclobutene, where sp^3 orbitals are used for the σ-bond. Rehybridization is necessary for product formation, and only p orbitals are shown in the transition state basis set for simplicity, and it does not affect the phase relationships. The dotted line in Figure 13.3(b), and subsequent diagrams, shows the phase relationships between adjacent, interacting orbitals. Note the correlation between the orbitals originally used to form the σ-bond. This represents a 'memory' of the original overlap.

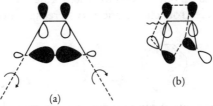

(a)

(b)

Figure 13.3 (a) Basis set for conrotatory cyclobutene ring opening; (b) Transition state phase relationships for cyclobutene conrotatory opening.

Examining the transition state for conrotatory ring opening (13.3(b)) we find one phase dislocation (phase dislocations only occur between different atomic orbitals). This is therefore an anti-Hückel system and has four electrons. *The conrotatory transition state for cyclobutene ring opening is therefore aromatic.* With a differently phased basis set, three phase dislocations would appear, but the conclusion remains unchanged.

A similar analysis for the disrotatory ring opening is given in Figure 13.4. The transition state in this example has two phase dislocations (zero or four with other basis sets). *The transition state for disrotatory cyclobutene ring opening is a four electron, anti-aromatic Hückel system.*

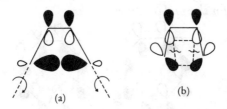

(a)

(b)

Figure 13.4 (a) Basis set for disrotatory cyclobutene ring opening. (b) Transition state phase relationships for cyclobutene opening.

It is clear that the two different ring-opening modes have different electronic, as well as stereochemical consequences. The implication of the analysis is that conrotatory ring opening, for cyclobutenes, has a lower energy transition state than disrotatory openings. This is quite consistent with the observed stereochemistry where conrotatory opening is the only path. In the language of the Woodward–Hoffmann rules thermal conrotatory ring openings in cyclobutenes is *allowed* and disrotatory opening is *forbidden*. These terms should not be taken in an absolute sense but more as an indication of which path, other things being equal, is likely to have the lowest energy, and therefore be preferred.

The Woodward–Hoffmann rules cannot predict whether a reaction is thermodynamically feasible as the analysis applies equally to both the forward and reverse reactions. If a reaction is feasible they predict whether it is likely to proceed at a measurable rate.

The more general Woodward–Hoffmann rules can be developed by examining another electrocyclic reaction. A stereospecific, electrocyclic ring closure is shown below,

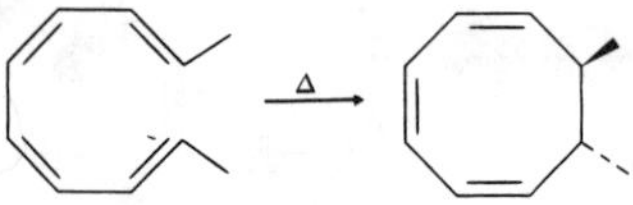

This reaction proceeds through a conrotatory ring closure. Figure 13.5 shows the transition state basis set. Using this basis set there is one phase dislocation, at the forming σ-bond. The transition state is an eight electron, aromatic, anti-Hückel system. Disrotatory ring closure for this reaction would proceed through an anti-aromatic eight-electron Hückel system, analogous to that shown in Figure 13.4(b). The alkene is drawn in the all-*s-cis* conformation. This is not extensively populated, but is the only conformation that leads to products.

The analysis of a number of systems allows the generalization, *thermal, electrocyclic reactions in the conrotatory mode are allowed for 4n electron transition states.*

The previous examples have been limited to thermal reactions but large numbers of photolytically induced, or *photochemical* reactions are known. *The Woodward–Hoffmann rules for photochemical reactions are always the reverse of those for thermal reactions.*

Figure 13.5 The basis set for an electrocyclic ring closure of an octatetraene, and the aromatic anti-Hückel transition state (right).

182

It is not so easy to explain why this should be so using our model but an adequate rationalization can be made. In a photochemical reaction an excited state is generated by the absorbance of light. The excited state is then directly converted to the product *ground* state. Calculations show that high energy excited states are most likely to be transformed into anti-aromatic transition states, leading to the opposite products to thermal reactions. The energy profiles in Figure 13.6 express this diagrammatically. The upper curve in Figure 13.6(a) represents the energy of the excited state and the lower curve the energy changes accompanying an allowed *thermal* reaction. When the excited state is populated, no reaction following the thermal path is likely. But Figure 13.6(b) shows the same excited state curve and the energy profile for the forbidden thermal route. Now, a populated excited state is very close in energy to a species leading to products. At the point where the two curves are closest in energy a 'crossover' from one state to the next can occur, and reaction proceeds, to give the thermally forbidden products.

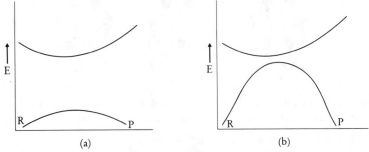

(a)   (b)

Figure 13.6   (a) The energy profile for an allowed thermal reaction R → P, and the energy of the photochemistry generated excited state. (b) The energy profile for a forbidden thermal reaction R → P approaches the energy of the photochemically generated excited state, allowing a crossover to give the thermally forbidden product.

An example of a photochemical $4n$-electron disrotatory ring closure is,

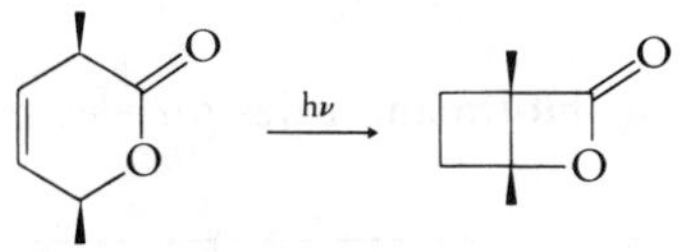

The eight-electron photochemical ring closure of substituted octatetraenes is also disrotatory and stereospecific.

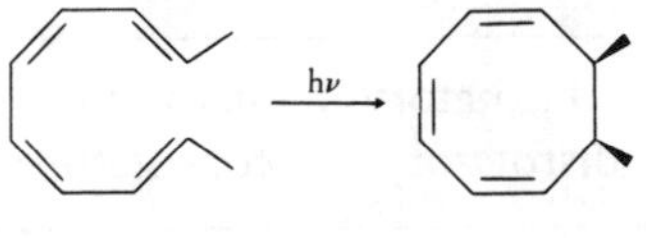

The product is the diastereoisomer of the thermal ring closure product.

The stereochemical implications of $4n+2$ electron electrocyclic reactions

183

only necessary to remember, or work out, one of the categories, as $4n$ and following reactions

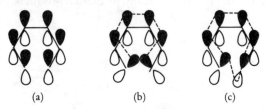

The transition states for the disrotatory and conrotatory closures are shown in Figures 13.7(b) and 13.7(c) respectively. The disrotatory transition state is a six-electron Hückel system and therefore *aromatic*. Conrotatory ring closure proceeds through an anti-aromatic, anti-Hückel transition state. It is therefore correctly predicted that thermal ring closure of substituted hexatrienes should be disrotatory, and the photochemical reaction should take the opposite, conrotatory pathway.

Figure 13.7   (a) The hexatriene basis set. (b) The transition state for hexatriene disrotatory ring closure. (c) The transition state for hexatriene conrotatory ring closure.

These results are generalized in the remaining Woodward–Hoffmann rules for electrocyclic reactions. *For $4n+2$ electron transition states, in electrocyclic reactions, the disrotatory mode is allowed for thermal reactions and the conrotatory mode for photochemical reactions.*

The Woodward–Hoffmann selection rules for electrocyclic reactions are summarized in Table 13.1.

Table 13.1. Woodward–Hoffmann rules for electro-cyclic reactions

Number of electrons in the cyclic transition state	Allowed, or preferred mode	
	thermal	photochemical
$4n$	conrotatory	disrotatory
$4n+2$	disrotatory	conrotatory

When the selection rules for pericyclic reactions are tabulated it is usually will now be examined. The transition state model can be applied to the $4n+2$ processes are always opposite, as are thermal and photochemical reactions.

13.4 Cycloaddition reactions

The method of analysing cycloaddition reactions by examining the transition states follows that for electrocyclic reactions. We shall compare the thermal reactions,

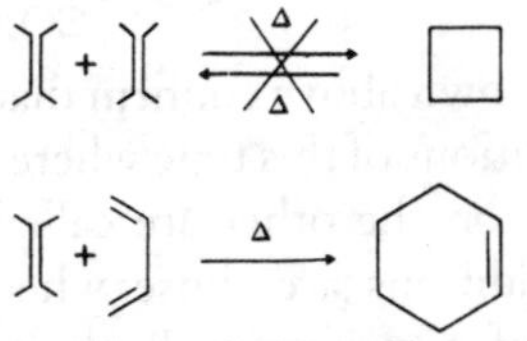

The transition state for ethene dimerization is shown in Figure 13.8. The two ethene molecules are assumed to approach in the simplest manner appropriate to σ-bond formation; along the principal axes of the p orbitals.

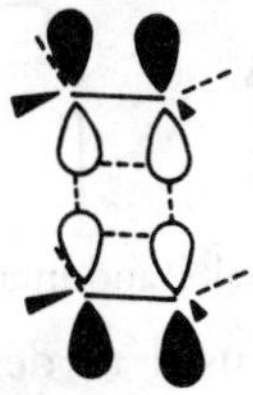

Figure 13.8 The phase relationships in the ethene dimerization transition state.

Figure 13.9 The phase relationships in the transition state for butadiene to ethene addition.

The transition state is a four electron Hückel system and therefore anti-aromatic. The thermal dimerization of ethene is therefore forbidden by the Woodward–Hoffmann rules. By contrast the addition of ethene to butadiene has an aromatic six-electron, Hückel transition state (Figure 13.9) and is therefore allowed thermally. Thermal cycloadditions of four π-electron systems to two π-electron systems are widespread and known collectively as *Diels–Alder* reactions. The four-electron system is called the *diene* and the two-electron fragment is the *dienophile*.

185

Like other pericyclic reactions Diels–Alder reactions are often stereospecific,

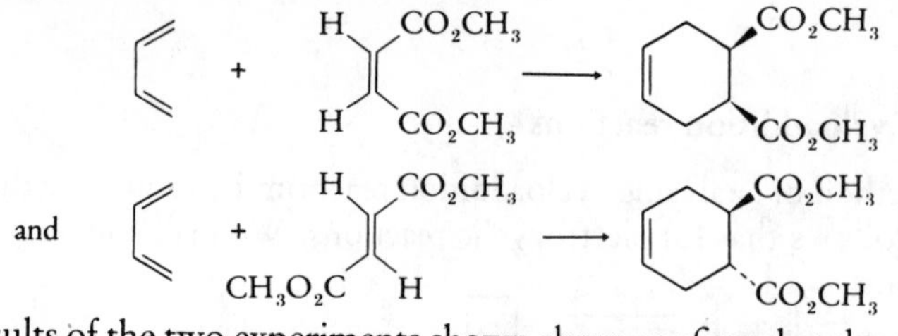

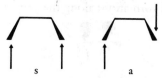

The results of the two experiments shown above confirm that the reactants approach as shown in Figure 13.9. Additions of this type where the termini of each π-system bond to the same face on the other are called *suprafacial–suprafacial* additions. *Antarafacial* additions are those where new sigma bonds are formed on *opposite* faces of a π-system. Both suprafacial and antarafacial modes of addition are shown in Figure 13.10.

Figure 13.10 Illustration of suprafacial(s) and antarafacial (a) addition to a π-system.

A particular nomenclature is used to describe cycloaddition reactions. Diels–Alder reactions are classed as $_4\pi_s + _2\pi_s$ cycloadditions. The numbers refer to the number of electrons *participating in the cyclic transition state*. The Greek symbols refer to the types of bond participating; both π-bonds in this example. The final subscripts describe the mode of addition, which for Diels–Alder reactions is suprafacial, s, for both components.

It is important when analysing pericyclic reactions to concentrate on the electrons directly involved in the transition state. Both of the following reactions are $_4\pi_s + _2\pi_s$ cycloadditions despite the conjugation with other π-bonds in the alkene fragments.

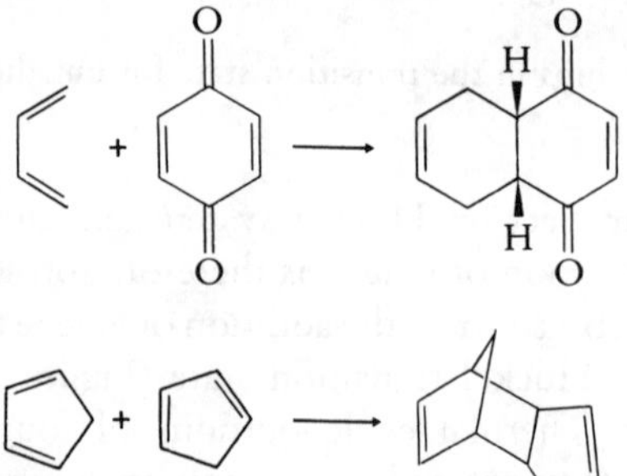

Diels–Alder reactions are only certain to be stereospecific when the regio-chemistry is unambiguous. Cycloadditions of unsymmetrically substituted

substrates are not necessarily regioselective,

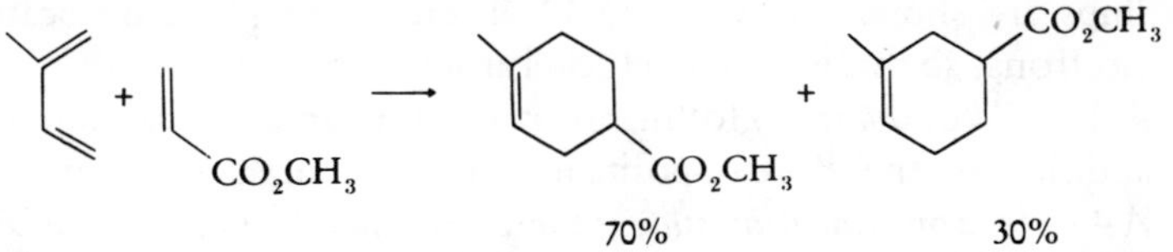

70%   30%

It is difficult to rationalize the observed regiochemistry of Diels–Alder reactions although some advanced analysis can be useful. These calculations are beyond the scope of this work.

The Woodward–Hoffmann rules for *suprafacial–suprafacial* cycloadditions can be summarized as *thermal cycloadditions are allowed for 4n+2 electron transition states and forbidden for 4n electron systems*. As before the photochemical selection rules are opposite to the thermal rules. *4n electron photochemical suprafacial–suprafacial cycloaddition reactions are allowed*.

The suprafacial–suprafacial photochemical addition of substituted ethenes is confirmed by the limited number of products obtained from the photolysis of butenes. Photolysis of Z-but-2-ene gives only two stereoisomers, both resulting from s–s addition.

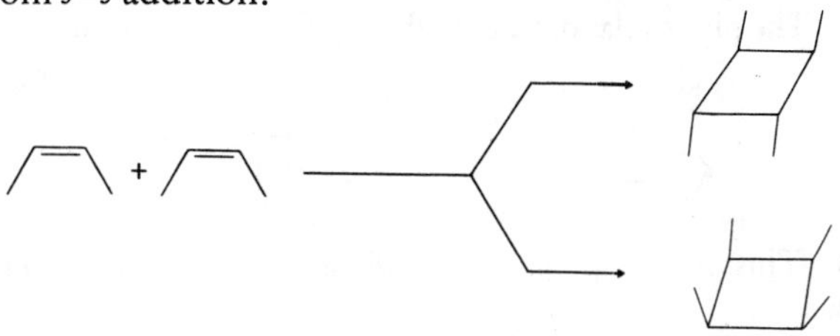

Antarafacial–suprafacial additions have also been observed in certain circumstances. The progress of a $_2\pi_a + _2\pi_s$ cycloaddition is shown in Figure 13.11. The essential stereochemistry can be reproduced by making models of alkenes and orientating them as shown in the figure. The two alkenes are initially arranged so that they are in perpendicular planes and one alkene lies across one face of the other. Only that face is accessible and the front lobes of its π-system start to bond to lobes on *opposite* faces of the second alkene. The bonding interaction becomes more pronounced as the alkenes twist about the axis joining the centres of both π-bonds. The transition state (13.11b) demonstrates, by means of the hatched lines how σ-bonding is developing on opposite faces of the forward alkene. The final product, 13.11(c), has the configuration retained at one alkene but the other has essentially undergone rotation about the double bond.

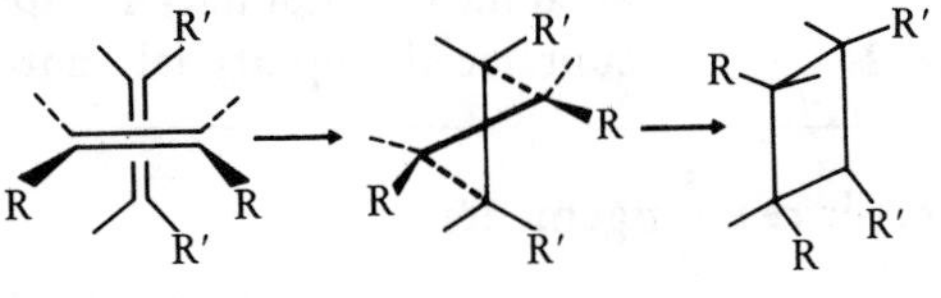

Figure 13.11   The progress of a $_2\pi_a + _2\pi_s$ cycloaddition.

187

The basis set for the transition state and the neighbouring phase relationships are shown in Figure 13.12. There is one phase dislocation and four electrons, so this is an anti-Hückel aromatic transition state.

The Woodward–Hoffmann rules for antarafacial–suprafacial cycloaddition are therefore opposite to those for suprafacial–suprafacial addition. *A 4n-electron transition state is therefore allowed for antarafacial–suprafacial cycloadditions.*

Antarafacial–suprafacial cycloaddition is too sterically hindered to be common but the highly strained alkene shown in Figure 13.13 spontaneously dimerizes in a $_2\pi_s+_2\pi_a$ cycloaddition.

Another example is the $_{14}\pi_a+_2\pi_s$ thermal cycloaddition shown in Figure 13.14.

The Woodward–Hoffmann rules for cycloaddition are summarized in Table 13.2.

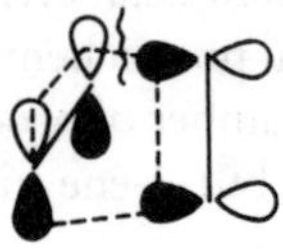

Figure 13.12   The phase relationships in the $_2\pi_a+_2\pi_s$ transition state.

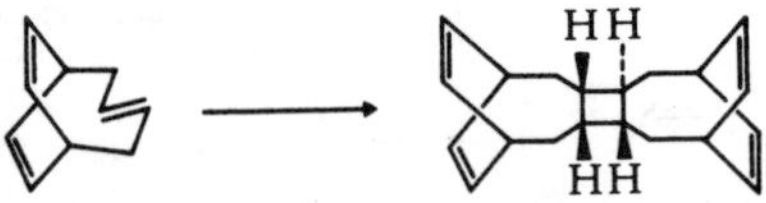

Figure 13.13   This alkene spontaneously dimerizes in $_2\pi_a+_2\pi_s$ manner to give the product shown.

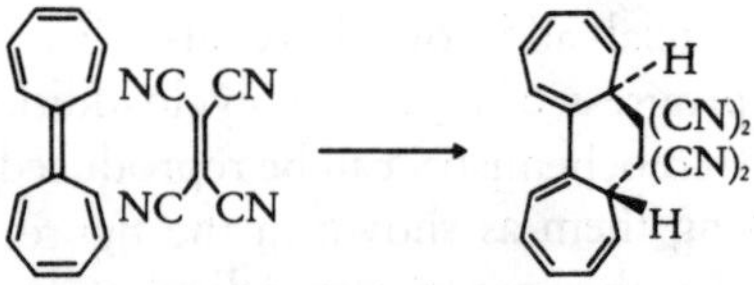

Figure 13.14   A $_{14}\pi_a+_2\pi_s$ thermal cycloaddition.

Table 13.2. Woodward–Hoffmann rules for cycloaddition reactions

	Allowed, or preferred mode	
Number of electrons	*thermal*	*photochemical*
4n	antarafacial–suprafacial	suprafacial–suprafacial
4n+2	suprafacial–suprafacial	antarafacial–suprafacial

13.5 Sigmatropic rearrangements

A sigmatropic change is one in which a σ-bond, flanked by one, or more,

188

π-electron systems, migrates to a new position in the molecule.

The *order* of the reaction $[i,j]$ is defined by the position of the new bond, relative to the original, and the number of electrons in the cyclic transition state. The termini of the original σ-bond are labelled 1,1 and the chains numbered sequentially.

$$i \ \overset{X^1}{\underset{\displaystyle \begin{array}{ccccc} 1 & 2 & 3 & 4 & 5 \end{array}}{\text{C—C=C—C=C}}}$$

$$i \ \overset{X^1}{\underset{\displaystyle \begin{array}{ccccc} 1 & 2 & 3 & 4 & 5 \end{array}}{\text{C=C—C—C=C}}}$$

After reaction, the termini of the new bond define the order. The reaction shown above is a [1,3] sigmatropic reaction.

Some reactions, such as the Cope reaction can involve two π-systems

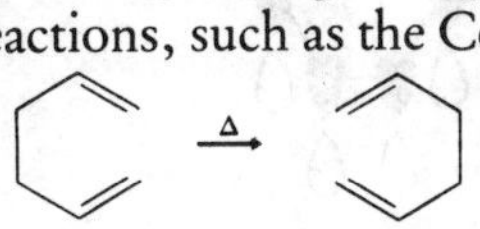

The Cope reaction is a 3,3 sigmatropic reaction

$$\begin{array}{cccc} & 1 & 2 & 3 \\ i & \text{C—C=C} \\ j & \text{C—C=C} \\ & 1 & 2 & 3 \end{array} \qquad \longrightarrow \qquad \begin{array}{ccc} 1 & 2 & 3 \\ \text{C=C—C} \\ \text{C=C—C} \\ 1 & 2 & 3 \end{array}$$

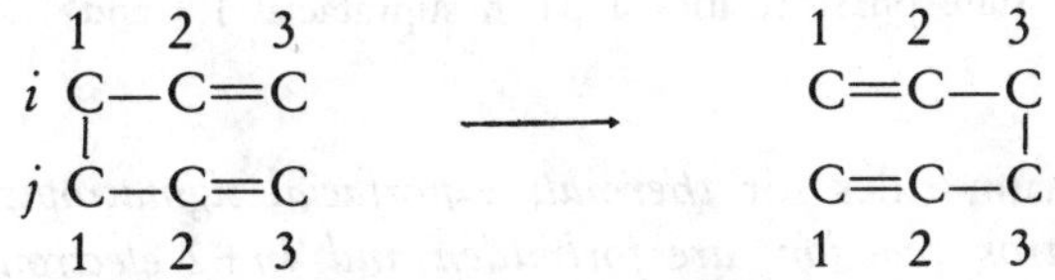

Migrations across a π-framework can be suprafacial or antarafacial as shown in Figure 13.15. In a suprafacial migration the migrating group remains associated with one face of the π-system throughout the reaction. A group that migrates from one face of the π-system to the other is said to undergo an antarafacial migration.

Figure 13.15 Suprafacial and antarafacial migrations across a π-ssytem.

It is simplest to set up the Woodward–Hoffmann rules for hydrogen migrations first. Consider the reaction below where the hydrogen atoms are 'scrambled' throughout the molecule.

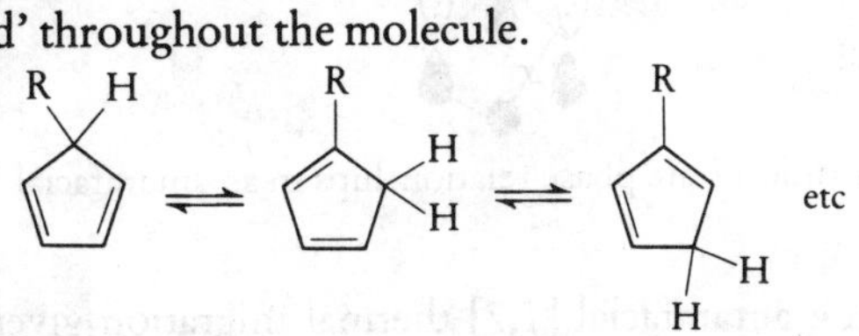

etc

The migration must be suprafacial for steric reasons but could possibly occur by [1,3] or [1,5] hydrogen shifts, or both. The [1,3] shift takes place by a migration across an allyl framework, and only two π-electrons are involved in the transition state. The [1,5] shift involves hydrogen migration to an *adjacent* atom but all four π-electrons are concerned in the transition state. Figure 13.16 gives the phase interactions in the basis set for [1,3] and [1,5] sigmatropic migrations. Both transition states have an even number of phase dislocations and are therefore Huckel systems. The [1,3] migration, being four-electron, is thermally forbidden and the [1,5] migration is a thermally allowed, six-electron aromatic transition state. Experiments confirm that the migration is indeed [1,5].

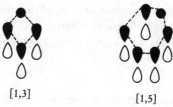

[1,3]   [1,5]

Figure 13.16 The transition state phase relationships in suprafacial 1,3 and 1,5 hydrogen migrations.

The Woodward–Hoffmann rules for *thermal, suprafacial sigmatropic reactions are that 4n electron reactions are forbidden and 4n+2 electron reactions are allowed. The opposite is true for photochemical reactions.*

An example of a suprafacial 4n-electron photochemical sigmatropic reaction is the [1,7] shifts in,

Antarafacial migrations have selection rules exactly opposite to those for suprafacial migrations.

An analysis of the transition state for the antarafacial [1,3] migration shown in Figure 13.17 illustates that the 4n electron systems are anti-Hückel aromatic and therefore allowed thermally.

Figure 13.17 The transition state phase relationships in an antarafacial 1,3 hydrogen migration.

The example of the antarafacial [1,7] thermal migration given in Figure 13.18 demonstrates that special steric factors are necessary for such migrations. A model of the starting material reveals that the π-system is twisted out-of-plane because of steric hindrance, so that the migrating hydrogen will always migrate from one face of the π-system to the other.

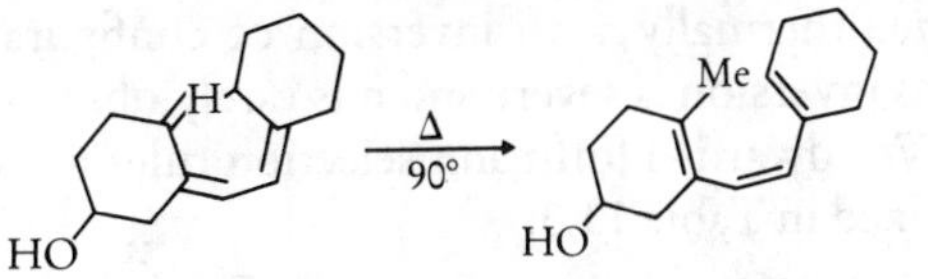

Figure 13.18   An antarafacial 1,7 thermal migration of hydrogen.

We now turn briefly to migrations of groups other than hydrogen. Two π-termini migrate in the stereospecific Cope reaction.

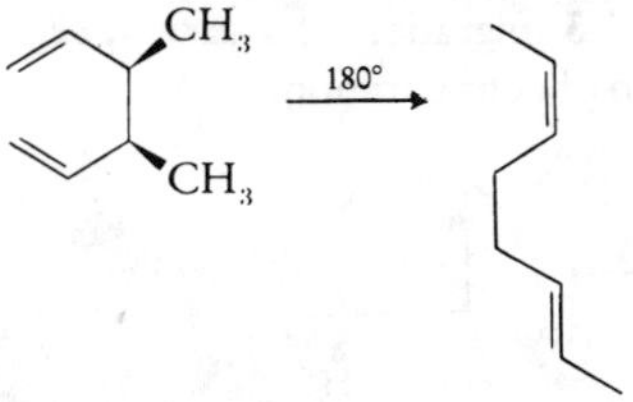

99.7%

The transition state is six electron, and assumed to take the low energy chair form as shown in Figure 13.19. A model is excellent for demonstrating how the chair leads directly to the observed E,Z isomer, and the unfavoured boat would give the E,E and Z,Z isomers. The Claisen reaction can be analysed similarly

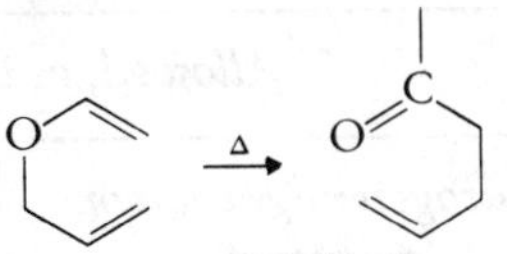

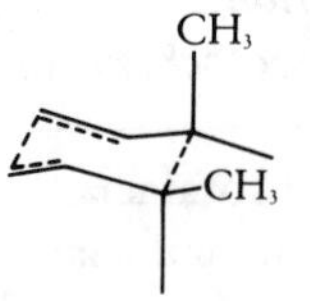

Figure 13.19   The favoured 'chair' arrangement for the Cope reaction.

There is one more important stereochemical feature of sigmatropic reactions that has not been discussed. When a saturated carbon centre migrates from one position to another, unlike hydrogen migrations, there are stereochemical consequences at the migration centre. The reaction can occur with retention or inversion of configuration at the migrating centre. Figure 13.20 compares the transition states for retention and inversion in a 1,3 suprafacial shift. The inversion transition state is a four electron anti-Hückel system and therefore aromatic.

The prediction is that sigmatropic reactions with inversion at the migrating

centre have selection rules opposite to those for hydrogen migration.

The elegant experiment shown in Figure 13.21 confirms that a 1,3 migration can take place thermally with inversion of configuration. In general, steric hindrance to inversion is severe and it is rarely observed.

The complex Woodward–Hoffmann selection rules for sigmatropic reactions are summarized in Table 13.3.

Figure 13.20   Phase relationships in the 1,3 migration of a carbon centre with (a) retention of configuration and (b) inversion of configuration.

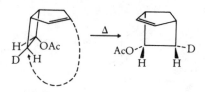

Figure 13.21   A 1,3 thermal migration with inversion of configuration at the migrating centre.

Table 13.3. Woodward–Hoffman rules for sigmatropic reactions

	Allowed, or preferred mode			
	hydrogen migration or retention		*inversion*	
Number of electrons	*thermal*	*photochemical*	*thermal*	*photochemical*
$4n$	antarafacial	suprafacial	suprafacial	antarafacial
$4n+2$	suprafacial	antarafacial	antarafacial	suprafacial

13.6 Summary of Sections 13.1–13.5

1. Pericyclic reactions are concerted, unaffected by catalysts or solvents, neither nucleophilic nor electrophilic reagents are involved, and have cyclic transition states.
2. The three classes studied are, electrocyclic reactions, in which a σ-bond is made between two termini of a conjugated π-system, or the reverse; cycloaddition reactions where two isolated π-systems react to form two new σ-bonds; and sigmatropic reactions in which a sigma-bond migrates across a π-system.

3. The Woodward–Hoffmann rules classify pericyclic reactions as 'allowed' or 'forbidden'. The allowed process is usually a low energy process and observed. The rules are developed here through the idea that 'aromatic' transition states are allowed and anti-aromatic transition states are forbidden for thermal reactions.

4. Aromatic transition states are those with $4n+2$ electrons in a Hückel array or $4n$ electrons in an anti-Hückel system. The basis set of atomic orbitals is examined for the transition state. The nature of the system is defined by the number of phase dislocations—zero or even for a Hückel system or odd for an anti-Hückel system.

5. The selection rules developed by examining transition states are given in Tables 13.1–13.3. Photochemical reactions always take opposite paths to thermal reactions. These rules are only for concerted reactions and relate to kinetic parameters not equilibrium data.

6. For electrocyclic reactions, two modes of ring opening or closure are recognized; conrotatory or disrotatory, defined in Figure 13.2.

7. Cycloaddition reactions can be suprafacial–suprafacial which is the stereochemically preferred mode, or suprafacial–antarafacial in exceptional circumstances. These terms are illustrated in Figures 13.10 and 13.11.

8. Sigmatropic reactions can have suprafacial or antarafacial migrations, as shown in Figure 13.15. Suprafacial migration is preferred although antarafacial migrations can occur with certain systems (Figure 13.18).

9. If a carbon-centred group migrates in a sigmatropic rearrangement it can do so with either retention or inversion of configurations. The selection rules are reversed for inversion of configuration, which is usually sterically disfavoured, but can occur (Figure 13.21).

Problems and Exercises

1. Only one product is obtained on thermolysis of *trans*-3,4-dimethylcyclobutene. Use models, and the Woodward–Hoffmann rules to predict which product is formed.

2. Explain how the following conversion can be achieved in two independent, concerted steps.

3. Addition of alkenes to 1,3 dipoles, such as diazomethanes, $> C = \overset{+}{N} = \overset{-}{N}$ $\longleftrightarrow > \overset{-}{C} - \overset{+}{N} \equiv N$ or nitrile oxides, $-C \equiv \overset{+}{N} - \overset{-}{O} \longleftrightarrow -\overset{-}{C} = \overset{+}{N} = O$ is common. By examining transition state basis sets decide whether the allowed thermal addition of diazomethane, H_2CN_2 to Z-but-2-ene is (a) a suprafacial–suprafacial or (b) antarafacial–suprafacial. Draw structures for all expected products.

4. Antarafacial–antarafacial cycloadditions are possible, in theory. Analyse the transition state for antarafacial–antarafacial addition of an alkene to a diene, $_2\pi_a + _4\pi_a$, and decide whether the reaction is thermally or photochemically allowed.

5. What product(s) would you expect from thermal 1,5 sigmatropic hydrogen migration in

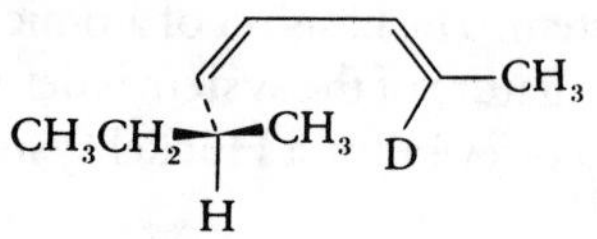

CHAPTER 14

Asymmetric synthesis— stereodifferentiation reactions

14.1 The creation of chirality in molecules

The reactivity of enantiomers in chemical reactions is always identical when there is no chiral agent present. But enantiomers can have widely different reactivities, especially in biological systems. As examples, only the L-amino acids can be used in protein synthesis, and R-N-isopropylnoradrenaline is more than 800 times more effective as a bronchochilator than its enantiomer. If we wish to study biological processes there is a need to find syntheses in which one enantiomer, from 2^n stereoisomers is produced.

In laboratory syntheses the overwhelming majority of readily available starting materials are achiral. Somewhere in the synthesis of a single enantiomer from an achiral molecule there needs to be a step in which one enantiomer is produced preferentially. That step is the subject of this Chapter.

In practice there are a number of ways of achieving the isolation of one enantiomer. A racemic mixture, as formed in achiral chemical reactions can be resolved. However, this separation requires three steps; a chemical modification to form diastereoisomers; a physical separation of the diastereoisomers and finally a reconversion of the individual diastereoisomers into the desired enantiomers. This process is time-consuming and economically disadvantageous. It is also aesthetically displeasing.

The ideal method for enantiomer production is to start with an achiral substrate, and under a chiral influence, synthesize one enantiomer selectively. The ideal has been achieved—the mention of enzymes can bring tears to the eyes of some organic chemists. An enzyme can catalyse a reaction so that one isomer is formed quantitatively, and, what is more it does so very rapidly in aqueous solution at about 30°C. Enymes are only found in biological systems and all efforts to design synthetic enzymes have been frustrated. An example of enzyme action, is,

$$\text{HO}-\underset{\underset{\displaystyle CH_2OH}{|}}{\overset{\overset{\displaystyle CH_2OH}{|}}{C}}-\text{H} \quad\xrightarrow[\text{enzyme}]{\text{ATP}}\quad$$

R-isomer only

Chemists are making use of enzymes in laboratory reactions with some success, but we shall not describe these complex reactions further.

Apart from the developing of enzyme reactions in laboratories, a huge amount of research is being done to try to develop other reactions for the synthesis of one enantiomer selectively. The subject of this Chapter is the development of some of these reactions, which in one case allows the preparation of 99.5% of one enantiomer.

The reactions fall into two classes. One is a set of reactions in which an achiral substrate gives an unequal amount of enantiomeric products. The other class has as a starting point, a chiral substrate that reacts with scalar reagents to produce unequal amounts of *diastereoisomers*, which can be separated and further modified.

Syntheses of both classes are called *asymmetric syntheses*, in which the creation of a new chiral element is achieved more or less selectively.

14.2 The nomenclature of asymmetric synthesis

14.2.1 Prostereoisomerism
Before describing asymmetric synthesis it is necessary to be able to specify the relationships of particular groups *within a molecule*. The concept of isomerism, as used in this work, has been confined to relationships between different molecules. Now it is essential to know how, and where, within a molecule, chirality can be produced.

It is impossible to produce a chiral centre in any XCH_3 molecule, by a single substitution. In the monochlorination of iodomethane, only achiral chloroiodomethane can be synthesized,

$$ICH_3 + Cl_2 \xrightarrow{h\nu} ICH_2Cl + HCl$$

An examination of the three hydrogen atoms in iodomethane in either a model or a projection formula, is convincing that each hydrogen is indistinguishable from the others. Such indistinguishable ligands are called *homotopic*. Topicity is the name given to relationships within molecules, and it is usually studied by the *substitution rule*. The substitution rule operates by examining the consequences of replacing, sequentially, two (or more) constitutionally identical ligands, by a test group that is different from all of the

196

other ligands. In iodomethane the substitution of any single hydrogen atom (by chlorine for example) gives an identical molecule to that produced on any other substitution. The hydrogen atoms themselves are therefore indistinguishable and called homotopic.

This analysis is continued by applying the substitution rule to chloroiodomethane (using bromine substitution as an example).

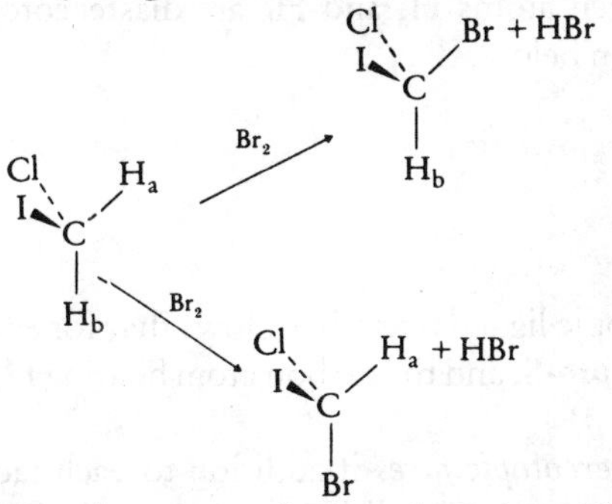

Here, the replacement of H_a and H_b has different stereochemical consequences, as enantiomers are formed on substitution. Ligands that produce enantiomers on application of the substitution rule are called *enantiotopic*. Enantiotopic ligands can be specified by a modified application of the sequence rules. The ligand to be labelled is arbitrarily assigned a higher priority than the other. The priority rules are then applied in the usual way, and a clockwise rotation demands that the ligand be labelled *pro*-R (sometimes H_R) and an anticlockwise rotation, *pro*-S. The application of these rules to ICH_2Cl gives H_a as *pro*-R (H_b—lowest priority; $I > Cl > H_a$) and H_b as *pro*-S. A carbon atom bearing enantiotopic groups is said to be *prochiral*.

The analysis of prochirality can also be extended to *faces* of molecules, if chirality can be generated by the addition to a face. If, say, acetaldehyde, is reduced by deuterium, enantiomeric alcohols are produced.

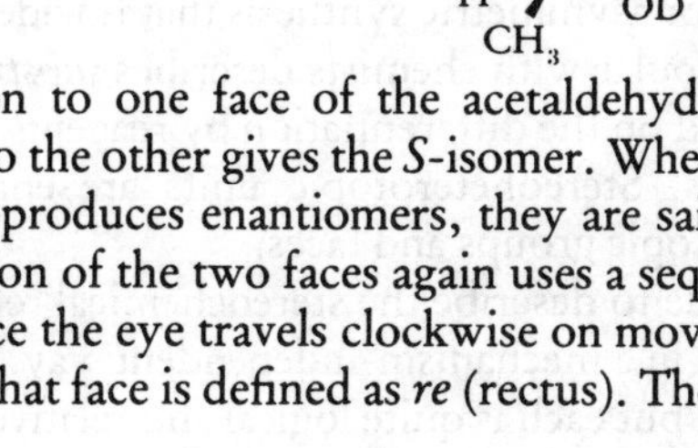

Addition to one face of the acetaldehyde gives *R*-deuterioethanol and addition to the other gives the *S*-isomer. When addition to opposite faces of a π-system produces enantiomers, they are said to be *enantiotopic faces*. The specification of the two faces again uses a sequence rule. If on looking down on one face the eye travels clockwise on moving from ligands to high to low priority, that face is defined as *re* (rectus). The opposite face is *si* (sinister).

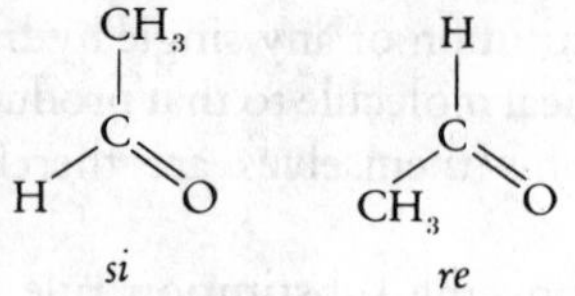

si   *re*

Ligands can also be diastereotopic, if application of the sequence rule leads to diastereoisomers. Hydrogen atoms H_a and H_b are diastereotopic in the R-1,2-dichloropropane shown below,

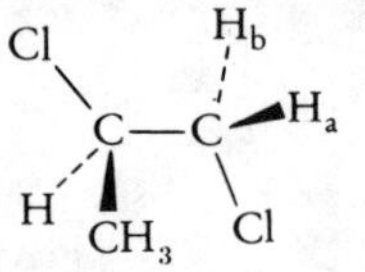

The labelling of diastereotopic ligands exactly follows that for enantiotopic groups. H_a is *pro-R* and H_b is *pro-S*, and the carbon atom bearing H_a and H_b is prochiral.

A molecule can have *diastereotopic faces* if addition to each face leads to diastereoisomers. The most common condition for a pair of diastereotopic faces is the presence of a prochiral sp^2 centre *and* a chiral centre in the same molecule.

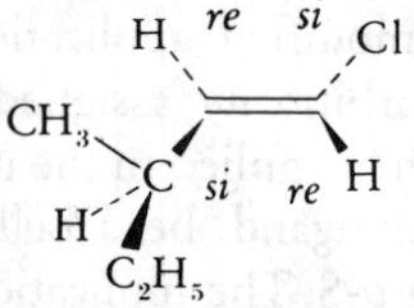

The molecule above has diastereotopic faces. In an example like this, each sp^2 prochiral centre is labelled separately as shown. The *syn* addition of deuterium to that molecule gives a pair of diastereoisomers,

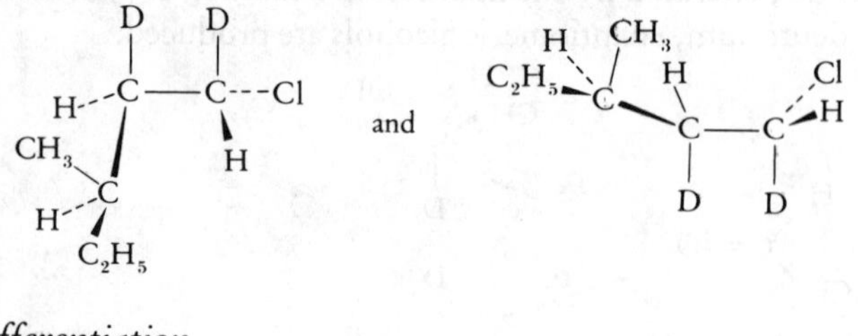

     and

14.2.2 Differentiation

A broad nomenclature system for asymmetric synthesis that is widely used in biology and becoming more popular with chemists describes *substrate specificity*. The nomenclature is based on the differentiation by reagents of stereoheterotopic units in substrates. Stereoheterotopic units are enantiotopic groups and faces, and diastereotopic groups and faces.

In this way an attempt is made to describe the stereochemical relationship between substrate and product, in a mechanism-independent way. There are six categories of differentiation, but each is quite logical and relatively simple

198

to remember. The classification falls into two classes, *enantiodifferentiation* and *diastereodifferentiation*.

In enantiodifferentiation, the chiral agent is external to the substrate, and enantiomers are usually formed. The three categories of enantiodifferentiation are,

1. *enantiotopos differentiation*, in which a chiral agent differentiates between enantiotopic ligands,

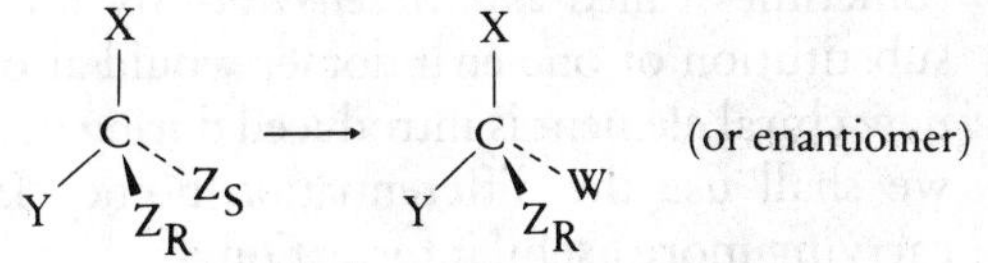

$$
\begin{array}{ccc}
\underset{Y}{\overset{X}{\underset{Z_R}{\overset{|}{C}}}}\!\!\!Z_S & \longrightarrow & \underset{Y}{\overset{X}{\underset{Z_R}{\overset{|}{C}}}}\!\!\!W \qquad \text{(or enantiomer)}
\end{array}
$$

2. *enantioface differentiation*, in which an enantiotopic face is discriminated by a chiral agent,

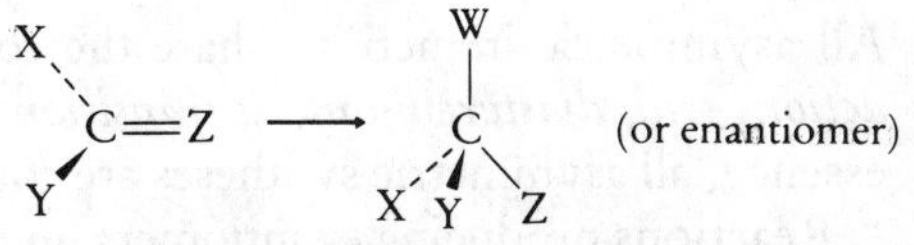

$$
\underset{Y}{\overset{X}{C}}\!\!=\!Z \longrightarrow \underset{X\ Y\ Z}{\overset{W}{C}} \qquad \text{(or enantiomer)}
$$

3. *enantiomer differentiation* does not strictly relate to asymmetric synthesis, but can be used for enantiomer separation,

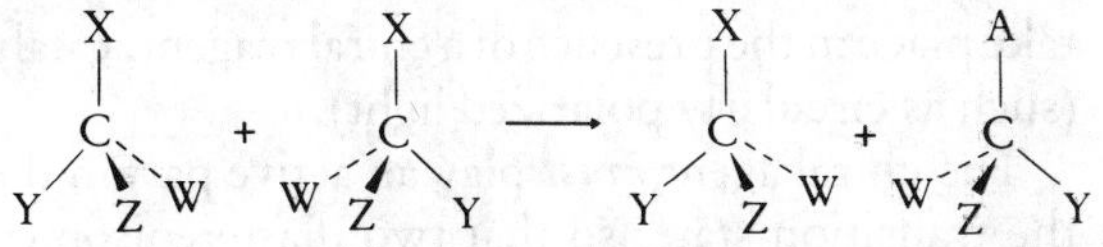

$$
\underset{Y\ \ Z}{\overset{X}{C}}\!W + \underset{W\ \ Z}{\overset{X}{C}}\!Y \longrightarrow \underset{Y\ \ Z}{\overset{X}{C}}\!W + \underset{W\ \ Z}{\overset{A}{C}}\!Y
$$

It is really a method of resolution—not creation of chirality.

In diastereodifferentiation the chirality is usually present *within the substrate*, and the reagent can be achiral. Diastereoisomers are the usual products. The classification of diastereodifferentiation follows that for enantiodifferentiation.

4. *diastereotopos differentiation* is where diastereotopic ligands are differentiated by the reagent,

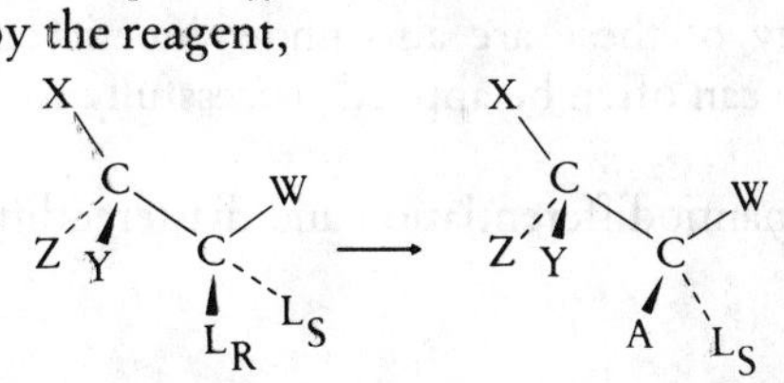

$$
\underset{Z\ Y}{\overset{X}{C}}\!\underset{L_R\ \ L_S}{\overset{|}{C}}\!W \longrightarrow \underset{Z\ Y}{\overset{X}{C}}\!\underset{A\ \ L_S}{\overset{|}{C}}\!W
$$

5. *diastereoface differentiation* in which diastereoisomers are produced by differential addition to π-systems,

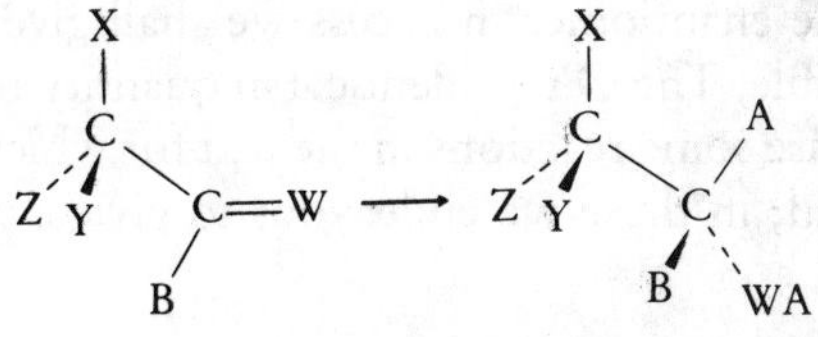

$$
\underset{Z\ Y}{\overset{X}{C}}\!\underset{B}{\overset{|}{C}}\!=\!W \longrightarrow \underset{Z\ Y}{\overset{X}{C}}\!\underset{B\ \ WA}{\overset{|}{C}}\!A
$$

6. *diasterereoisomer differentiation* reactions are those in which diastereo-isomers react at different rates. In fact diastereoisomers have different internal energies and should, in principle, react at different rates—even when the product is common. Once again this category is included for completeness and will not be discussed further.

Other nomenclature systems for asymmetric synthesis are not so informative, and can be ambiguous. Reactions producing one enantiomer in excess are sometimes called *enantioselective*. By this definition the stereospecific S_N2 substitution of one enantiomer would also be enantioselective, although no new chiral element is introduced during the reaction. Where ambiguity exists we shall use the differentiation nomenclature as being more precise and carrying more useful information.

14.3 The principle of asymmetric syntheses

All asymmetric inductions share the common principle *that competing reactions with diastereoisomeric transition states take place at different rates*. In essence, all asymmetric syntheses are competing reactions.

Reactions producing enantiomers, in the *absence* of a chiral influence, have transition states of identical energies and therefore take place at identical rates to give equal amounts of the enantiomers. Enantiodifferentiation reactions take place in the presence of a chiral reagent, catalyst, solvent or physical force (such as circularly polarized light).

The chiral agent *must* play an active part in the reaction and be integral to the transition state, so that two diastereoisomeric transition states are produced. One stereoisomer is therefore produced more rapidly than the other, in an asymmetric synthesis.

All reactions in which one enantiomer is formed in excess from achiral substrates *must* be under kinetic control. Thermodynamic control in such cases can only lead to racemic mixtures. Diastereodifferentiation reactions would produce unequal amounts of diastereoisomers, even by thermo-dynamic control, but many of these are also under kinetic control. The Curtin–Hammett principle can often be applied successfully to asymmetric syntheses.

We shall now examine enantiodifferentiation and diastereodifferentiation separately.

14.4 Enantiodifferentiation

14.4.1 Enantiotopos differentiation

In reactions generating one enantiomer in excess we shall give the *Optical Yield* (OY) wherever possible. The OY is identical in quantity to the optical purity (Chapter 8). There are some reactions in the text for which the optical yield has not been reported; in these we endeavour to give as much useful information as possible.

The number of enantiotopos differentiations, outside enzyme chemistry, is rather small, compared with face-differentiations in general. They are, however, known and raise some interesting points.

One example where the enantiotopic groups are CH_2 groups is,

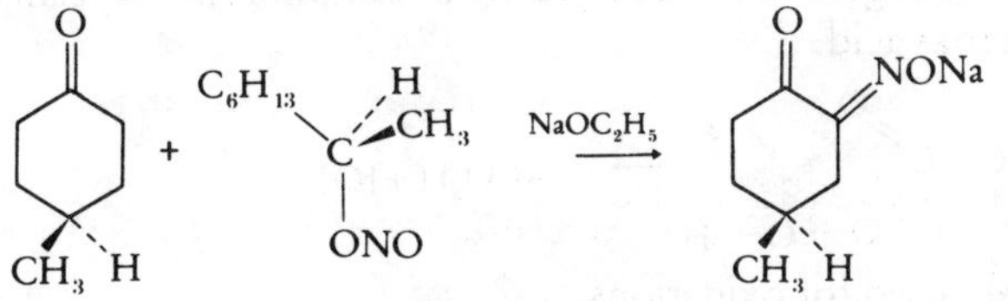

The optical yield, at less than 5% is low, but the chiral centre (C-4) is rather remote from the reaction site. The mechanism of this reaction is quite complex and we are unable to rationalize the observed product formation. In face-differentiations the mechanisms have often been well studied and are well enough understood to allow the prediction of the outcome of analogous reactions. The particular reaction shown above suffers from twin disadvantages. The first, the poor OY is obvious and the second concerns the wastefulness of using one mole of chiral reagent in the generation of one mole of product. Ideally, enantiodifferentiation reactions should use chiral catalysts, in trace amounts. In this way an economic asymmetric synthesis could be developed.

A small optical yield has been reported for the catalytic dehydration of *trans*-4-methylcyclohexanol, using a chiral acid as catalyst,

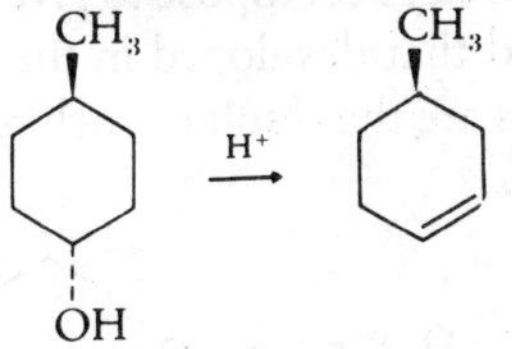

There are several sets of enantiotopic hydrogen atoms in the cyclohexyl-alcohol substrate, as shown in Figure 14.1. Enantiotopic pairs are labelled with the same letter. Only H_a and H'_a are antiperiplanar to the hydroxy group and it is differentiation of these that gives rise to the chiral alkene. The chiral acid used in this synthesis is a derivative of camphor, an abundant naturally-occurring chiral molecule. Because of its availability camphor and its derivatives are widely used in the preparation of chiral reagents and catalysts.

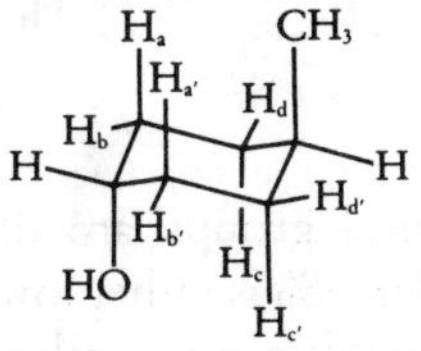

Figure 14.1   Enantiotopic hydrogen atoms in *trans*-4-methylcyclohexanol.

201

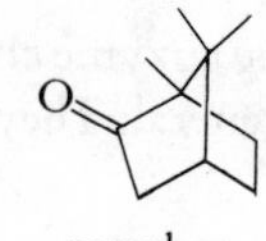

camphor

An oxidizing agent that is a derivative of camphor, peroxycamphoric acid is a typical peroxyacid,

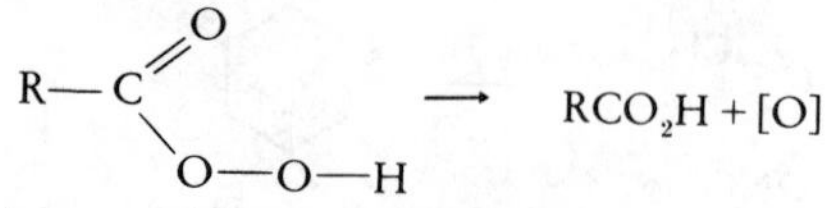

$$R—C\overset{O}{\underset{O—O—H}{}} \quad \longrightarrow \quad RCO_2H + [O]$$

and is widely used for oxidations.

Although reactions of peroxyacids are not catalytic, the acid is recovered after reaction and can be reconverted, quite simply, to the peroxyacid. An interesting use of peroxycamphoric acid is the following reaction, in which enantiotopic lone pairs of electrons are differentiated, with an optical yield of 5.4%.

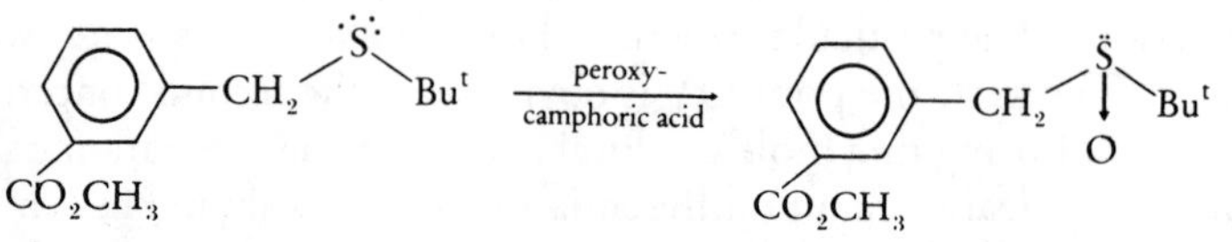

Enantiomers are not always formed in enantiodifferentiation reactions. In the examples shown above the diastereoisomeric transition states decompose to yield two chiral fragments, one of which is the product shown. Under different circumstances the transition state can decompose to give a product in which the chirality from the reagent, and that developed in the reaction are both present. Diastereoisomeric products are the result of such reactions.

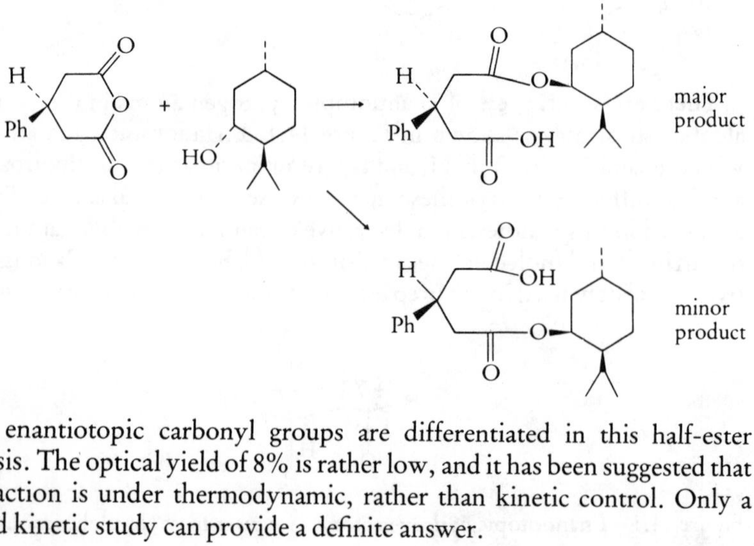

major product

minor product

The enantiotopic carbonyl groups are differentiated in this half-ester synthesis. The optical yield of 8% is rather low, and it has been suggested that this reaction is under thermodynamic, rather than kinetic control. Only a detailed kinetic study can provide a definite answer.

202

14.4.2 Enantioface differentiation

Currently, enantioface differentiations are more effective than enantiotopos differentiations. There are also useful models for the transition states of many enantioface differentiations.

A good example is the use of sterically-hindered Grignard reagents to reduce carbonyl compounds, to alcohols. If the Grignard reagent is chiral, then one enantiomer of the alcohol can be formed in excess, through enantioface differentiation.

$$PhCOPr^i + S - ClMgCH_2CH(C_2H_5)CH_3 \rightarrow S - Ph(Pr^i)CHOH$$
$$24\% \text{ OY}$$

The steric bulk of the Grignard reagent can affect the optical yield.

$$PhCOPr^i + S - ClMgCH_2CH(C_2H_5)Ph \rightarrow S - Ph(Pr^i)CHOH$$
$$82\% \text{ OY}$$

The results of these reactions have been rationalized by examining models of the transition states. It is postulated that the predominant product will be the one derived from the *least-hindered* transition state. The supposed geometries for the transition states for reduction of ketones by Grignard reagents is shown in Figure 14.2. The ligands are labelled according to their steric bulk, R_L and $R_{L'}$ are the larger ligands and R_S and $R_{S'}$ are the smaller. The transition state is six-centred, with the magnesium atom and hydrogen atom placed for near-simultaneous transfer to the termini of the carbonyl group. The diastereoisomeric transition states **1** or **2** are formed, according to which face of the alkene is attacked by the Grignard reagent. Transition state **1** is assumed to be lower in energy, as the two bulky groups R_L and $R_{L'}$ are well-separated.

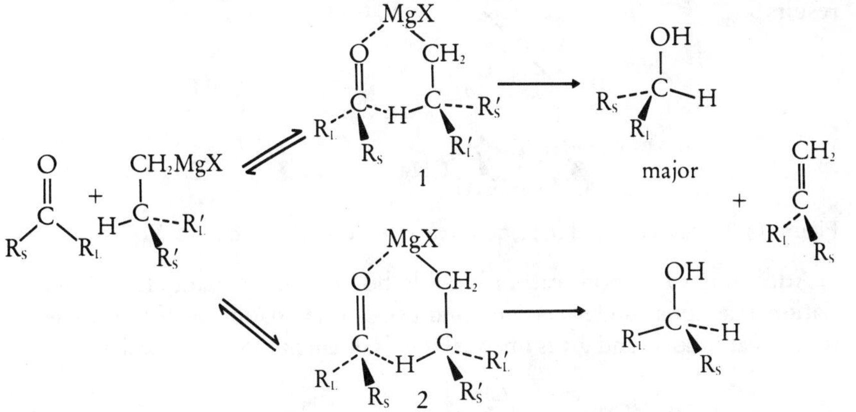

Figure 14.2 Transition states for the reduction of ketones with chiral Grignard reagents.

In transition state **2** there is more crowding, owing to the proximity of the larger ligands. On the assumption that $Ph > Pr^i$, $Ph > Et$ and $Et > Me$ the configuration of the chiral alcohols formed by Grignard reagent reductions of ketones can be explained. Although this transition state model is effective for many reactions there are a number of occasions where it does break down. Firstly, the model is oversimplified and takes no account of the reaction conditions, such as solvent and temperature, that profoundly alter the structure of Grignard reagents. Secondly, it is not always simple to decide which are the larger groups. Generally the order of decreasing size is,

$$Ph > Bu^t > Pr^i > Et > Me > H$$

Problems do arise with the phenyl group. In some reactions it appears to be larger than the tertiary butyl group and in others it behaves as if it were smaller. Despite this, for reactions without tertiary butyl groups, and with simple alkyl groups the model can be used with a reasonable degree of confidence. It would not be wise, though, to assign absolute configurations on the results of this model alone.

A closely-related reaction to the Grignard reagent reductions is the Meerwein–Ponndorf–Verley reaction in which an aluminium salt catalyses the reaction,

$$RR'C{=}O + R''R'''CHOH \rightleftharpoons RR'CHOH + R''R'''C{=}O$$

The favoured transition state for the reaction between S-Pr^iMeCHOH and $C_6H_{11}COCH_3$ is shown in Figure 14.3. The optical yield in this example is 22%. It is often found that the highest optical yields in Meerwein–Ponndorf–Verley reactions are at low percentage conversions of substrate. As the reaction proceeds the chiral alcohols can be racemized under the reaction conditions. Ultimately thermodynamic control–complete racemization–results.

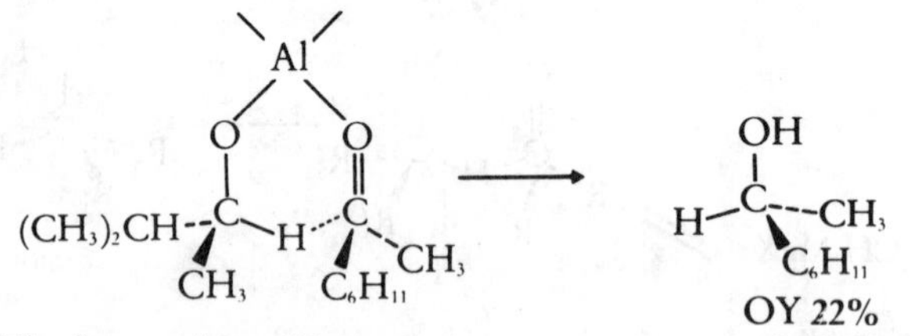

Figure 14.3 The favoured transition state in a Meerwein–Ponndorf–Verley reaction.

Additions to carbon–carbon double bonds can be enantioface differentiation reactions, under certain conditions. Epoxidation of E-but-2-ene by peroxycamphoric and gives unequal yields of enantiomeric epoxides.

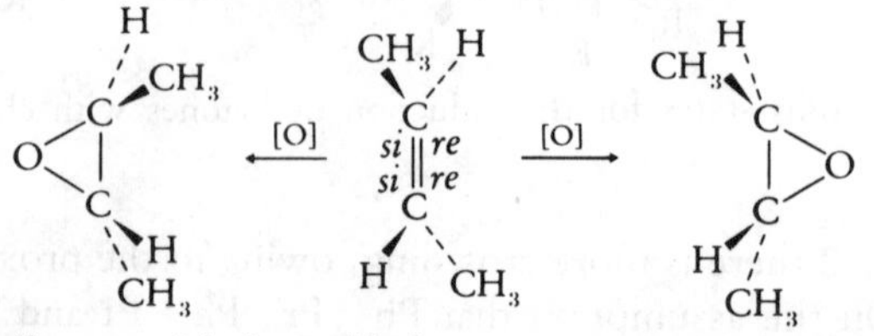

But a similar epoxidation of Z-but-2-ene can only give the *meso* epoxide.

204

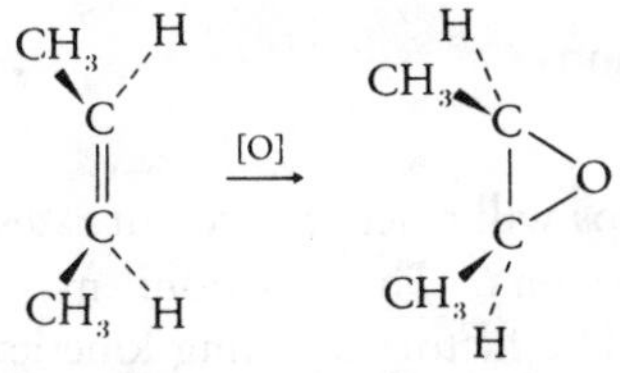

[Structure diagram at top showing alkene to epoxide conversion with [O]]

Enantioface differentiation of symmetrically substituted Z-alkenes can be demonstrated by the unsymmetrical addition of chiral reagents. An 81% optical yield has been obtained for the addition of dipinnanylborane to Z-but-2-ene. The reaction scheme is shown in Figure 14.4. The borane addition product shown in the figure is the diastereoisomer formed in excess, and it can be stereoselectively converted to the alcohol shown. Differentiation of the enantiofaces of propene is not so pronounced, as the borane adds regioselectivity to the CH_2 carbon atom, which is more remote from the prochiral centre.

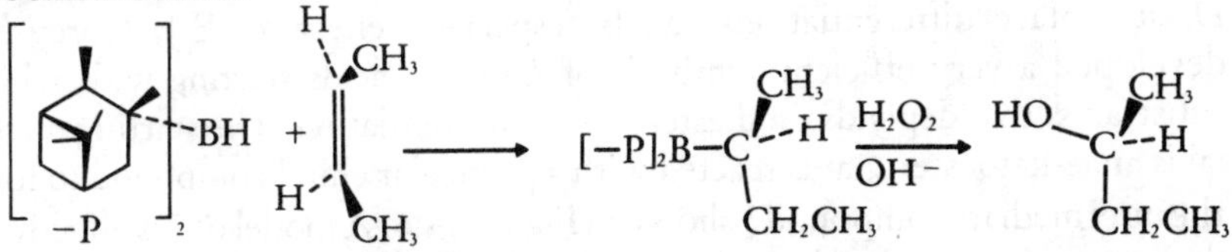

[Reaction scheme with borane structure]

$$[\text{structure}]-BH + \text{alkene} \longrightarrow [-P]_2B-C\overset{CH_3}{\underset{CH_2CH_3}{|}}-H \xrightarrow[OH^-]{H_2O_2} HO-C\overset{CH_3}{\underset{CH_2CH_3}{|}}-H$$

Figure 14.4 Enantioface differentiation by a chiral borane.

One of the most exciting areas of asymmetric synthesis is the use of transition metal catalysts as differentiation agents. The catalysis can be on surfaces (heterogeneous) or by metal complexes in solution (homogeneous). One of the most successful homogeneous catalysts for hydrogenation reactions is Wilkinson's Catalyst $(Ph_3P)_3RhCl$, named for its discoverer Sir Geoffrey Wilkinson (Nobel Laureate in Chemistry 1973). Recently Wilkinson-type catalysts with chiral phosphine ligands have been used with great success in enantioface differentiation reactions. The anti-Parkinson's disease drug L-dopa has been synthesized in 95% optical yield using this type of catalyst. The reaction is shown in Figure 14.5. Mechanistically, these reactions are complex, but well-understood. The differentiation step involves enantioface differentiation in the coordination of the alkene onto the metal.

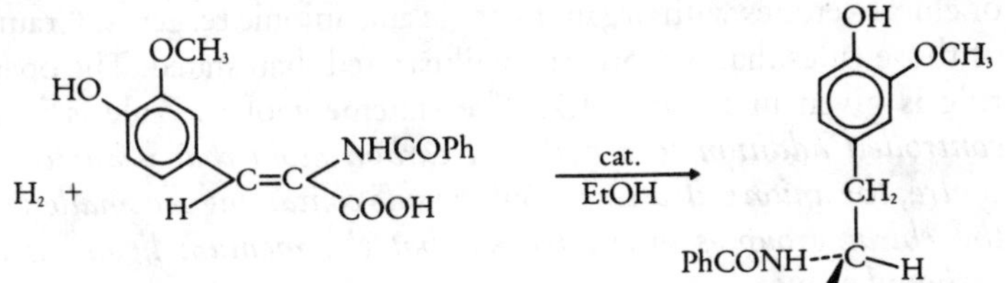

[Reaction scheme for L-DOPA synthesis]

$$H_2 + \text{substrate} \xrightarrow[EtOH]{cat.} \text{L-DOPA}$$

OY 96%
L-DOPA

Figure 14.5 Highly selective enantioface differentiation in the presence of a chiral Wilkinson-type catalyst.

14.5 Diastereodifferentiation reactions

14.5.1 Diastereotopos differentiation

In principle, any diastereotopic groups will react at different rates to give an unequal distribution of diastereoisomers. The reactions may be under thermodynamic or kinetic control. The factors affecting kinetic control in diastereotopos differentiations have not received overwhelming attention and we limit this section to one example,

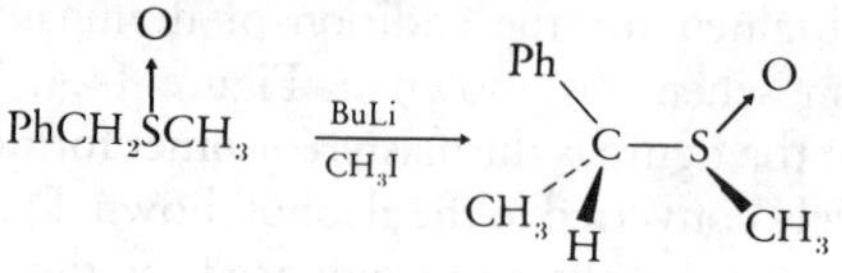

30% excess

14.5.2 Diastereoface differentiation

Diastereoface differentiations can be especially effective. E.J. Corey has developed a very efficient synthesis of L-amino acids starting with achiral substrates, that depends on diastereoface differentiation. The starting material is an α-ketoester that is reacted with a particular chiral compound to form the intermediate compound 3 shown in Figure 14.6. A model of 3 is effective in demonstrating restricted access to one of the C=N faces. Corey found that hydrogenation, followed by hydrolysis gave 96–99% optical yields of the L-α-amino acids, through diastereoface differentiation.

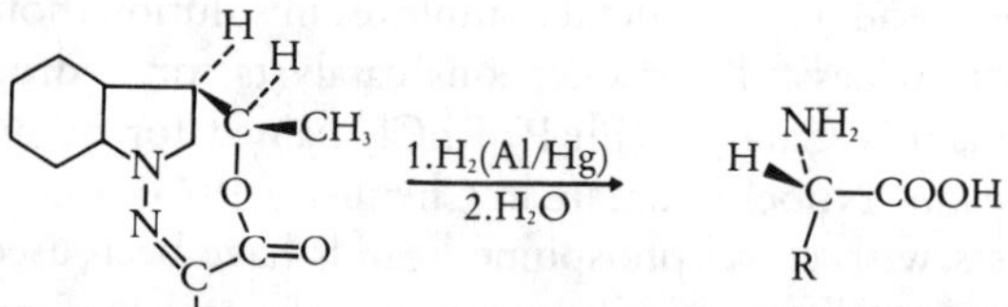

96–99% OY

Figure 14.6 The Corey synthesis of L-amino acids, by diastereoface differentiation.

In the 1950's D.J. Cram formulated a rule, now known as *Cram's rule*, to account for the preferential formation of one diastereoisomer in the reaction of chiral ketones with organometallic and anionic reagents. Cram's rule is one of those rules that is more easily illustrated than stated. The operation of the rule is given in Figure 14.7. The statement of the rule is *in a kinetically controlled addition to a carbonyl carbon atom that is adjacent to a chiral centre, the anion will attack from the side containing the smallest ligand, when the chiral group is orientated so that the medium ligand is eclipsing the carbonyl group.*

Some applications of Cram's rule are shown in Figure 14.8. It is observed that the degree of diastereoface differentiation increases as the size of the attacking group increases, which is consistent with the model.

The rule holds well for alkyl and phenyl groups, but has to be modified when one of the ligands on the chiral carbon atom is highly polar. In the case of ligands like hydroxyl and alkoxyl, with a high coordinating ability with metals, that ligand is placed to eclipse the carbonyl group. The reagent organolithium or Grignard reagent then bridges between that ligand and the carbonyl group, in the manner shown in Figure 14.9. The coordinated ligand R'' subsequently attacks from the least hindered side.

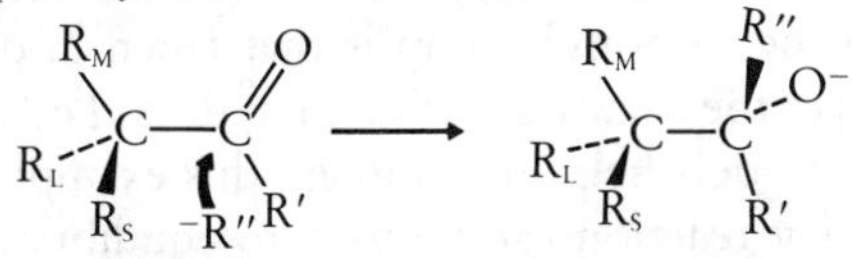

Figure 14.7   Illustration of Cram's rule.

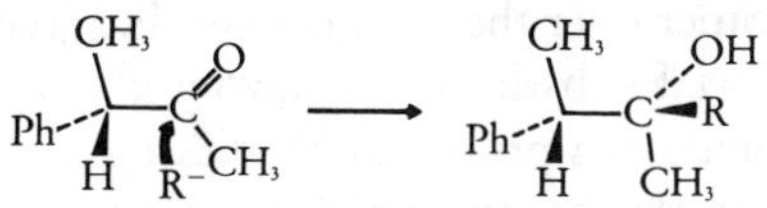

Figure 14.8   Application of Cram's rule. $R = CH_3$, $OY = 66\%$; $R = C_2H_5$, $OY = 75\%$; $R = Ph$, $OY = 83\%$.

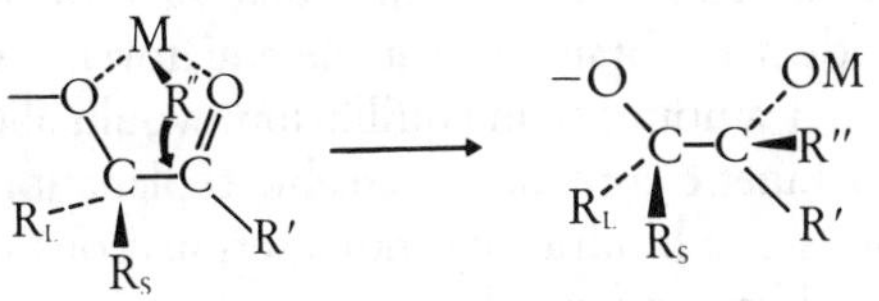

Figure 14.9   The modified Cram's rule.

At about the same time as Cram's rule was developed, V. Prelog (Nobel Price 1975) formulated a similar rule (*Prelog's rule*) for the addition of organometallic reagents to the chiral alcohol ester of an α-ketoacid. The reaction initially produced the ester of an α-hydroxy acid which yields the α-hydroxy acid on hydrolysis. The operation of Prelog's rule is shown in Figure 14.10. The diastereoface differentiation is not so great as in the Cram's rule examples, but the chiral influence in this example is much further removed from the carbonyl group.

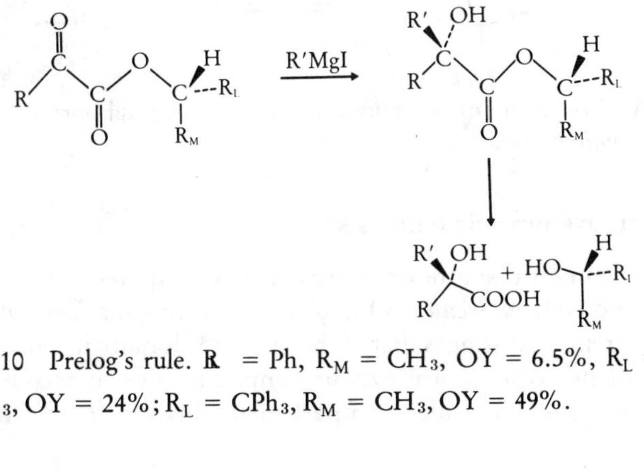

Figure 14.10   Prelog's rule. $R = Ph$, $R_M = CH_3$, $OY = 6.5\%$, $R_L = C(CH_3)_3$ $R_M = CH_3$, $OY = 24\%$; $R_L = CPh_3$, $R_M = CH_3$, $OY = 49\%$.

The success of Cram's rule, Prelog's rule and other similar transition state models rests on the assumption that the transition states have considerable *reactant character*. In other words it is valid to examine the steric effects on bringing two reactants together on the assumption that the *initial* steric interactions dominate. This assumption does usually hold but the examples of diastereoface differentiation shown in Figure 14.11 provide a warning! The least-hindered approach to the carbonyl group is that taken in diisobutyl-borane reduction, which gives the axially substituted hydroxyl compound in 96% yield. (You may find models help in following this example). Now a Meerwein––Ponndorf–Verley reaction can be used to equilibrate the axial cyclohexanol with the equatorial diastereoisomer. This equilibration results in a 4:1 excess of equatorial isomer over the axial isomer, in agreement with the qualitative idea that equatorial substitution is favoured. The axial diastereoisomer *must* therefore have been produced by diastereoface differentiation under kinetic control, in the borane reduction. If we examine the reduction of 4-*t*-butylcyclohexanone with lithium aluminium hydride (LAH), which is not a hindered reagent we find that 95% of the thermodynamically favoured product is obtained. But the equatorially substituted alcohol is formed in a greater amount than equilibrium would allow, so this too must be formed under kinetic control. A sensible explanation for this is that in the LAH reduction, steric hindrance is not a serious constraint. If the transition state is product-like, and much lower in energy for the route leading to equatorial hydroxyl substitution then that diastereoisomer will dominate by kinetic control. This rather complex example illustrates the fact that generalizations are very useful, but only when their limits are appreciated!

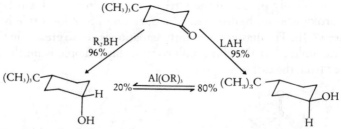

Figure 14.11 Reduction of 4-*t*-butylcyclohexanone by diisopropylborane and lithium aluminium hydride.

14.6 Absolute asymmetric synthesis

All of the asymmetric syntheses described have required the presence of chiral, and resolved, *molecules*. One of the problems that several research groups are currently tackling is that of the origin of chirality in living systems. A coherent theory of chemical evolution must be able to account for the production of excess molecules of a particular chirality, from which living systems could develop.

Obviously the use of chiral molecules cannot be postulated, as that engenders a 'chicken and egg' paradox. Two theories seem to be popular. The first is that of 'accidental' crystallization of one enantiomer, constituting a *de facto* resolution. This kind of process can be reproduced in a laboratory. For example when a racemic solution of $Ph(C_2H_5)(CH_3)\overset{+}{N}CH_2CH=CH_2 I^-$ is left to stand for long periods, one enantiomer, sometimes *R*, sometimes *S*, will spontaneously crystallize. Such 'spontaneous' crystallization can be induced by chiral particles of an achiral substance. Both glycine and quartz (SiO_2) have chiral crystals.

The second process by which enantiomers could be selectively produced, in the absence of other chiral molecules, is a chemical reaction brought about in the presence of a chiral force, such as circularly polarized light. It is known that light, reflected from the surface of water is weakly polarized. There have been numerous attempts to use right and left circularly polarized light to induce asymmetric syntheses. The optical yields in such reactions have so far been minute, but the encouraging observation is that the rotations of products from light of opposite polarizations is usually approximately equal and opposite. One of the more successful of these absolute syntheses is shown in Figure 14.12. The hexahelicene formed has a specific rotation of $-7.5°$ from right circularly polarized light and $+7.5°$ from left circularly polarized light.

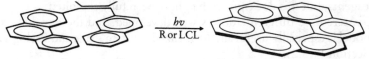

Figure 14.12   Absolute asymmetric synthesis of hexahelicene, using RCL or LCL.

14.7 Summary of Sections 14.1–14.6

1. Resolved, or partially resolved chiral compounds can only be prepared in the presence of a chiral influence.
2. An asymmetric synthesis is one in which a new chiral unit is formed selectively.
3. Chirality can be developed at *prochiral centres*. A prochiral centre is one bearing *enantiotopic* or *diastereotopic* ligands, or is part of a π-system with *enantiotopic faces* or *diastereotopic faces*.
4. The substitution rule is used to determine the presence of stereoheterotopic ligands. If a pair of constitutionally identical ligands is substituted sequentially and enantiomers are formed, these ligands are enantiotopic. If diastereoisomers are produced the ligands are diastereotopic. If identical compounds are formed they are homotopic. The presence of stereoheterotopic faces is determined by a symmetrical addition across the face and the resulting compounds analyzed as above.
5. Modified sequence rules are used to specify stereoheterotopic units (Section 14.2.1).

6. Product asymmetry in reactions can be generated in six ways, depending on the ability of the reagent to differentiate different units of prochirality or chirality. The differentiation reactions are (Section 14.2.2): enantiotopos differentiation; enantioface differentiation; enantiomer differentiation; diastereotopos differentiation; diastereoface differentiation; diastereoisomer differentiation.

7. Enantiomer and diastereoisomer differentiation are *resolutions* not asymmetric syntheses and were not studied further in this Chapter.

8. All asymmetric syntheses are competition reactions proceeding through diastereoisomeric transition states. Kinetic control is essential in enantiomer production.

9. Enantiotopos differentiation (Section 14.4.1) are not especially useful, except when enzymes are the chiral influence.

10. Enantioface differentiation (Section 14.4.2) can be very effective. A transition state model for Grignard reagent reductions and Meerwein–Ponndorf–Verley reactions is useful, and assumes that the least hindered transition state has the lower energy.

11. Transition metal catalysts, such as modified Wilkinson's catalysts can be effective in asymmetric inductions and do not require expensive stoichiometric amounts of catalyst.

12. Diastereotopos differentiation should always proceed, even with achiral reagents, but this does not seem to have been fully exploited.

13. Diastereoface differentiation is well developed, and the transition state model is especially useful and formalized in Cram's rule and Prelog's rule (Section 14.5.2).

14. All transition state models assume that the transition state is reactant-like. This is usually valid for hindered systems but care must be observed in less hindered reactions, where product-like transitions states may dominate.

15. Absolute asymmetric syntheses are those in which a chiral force is the chiral agent. These are not efficient in laboratories, but experiments with circularly polarized light have been partially successful.

Problems and Exercises

1. Use the substitution rule, and models where necessary, to help you to locate enantiotopic or diastereotopic groups or faces in the following compounds:

2. Classify the following reactions according to the differentiation nomenclature, and predict the configuration of the major products.

(a) $PhCDO + R(CH_3)_3C(CH_3)CHOH$ in the presence of an aluminium salt.

(b)

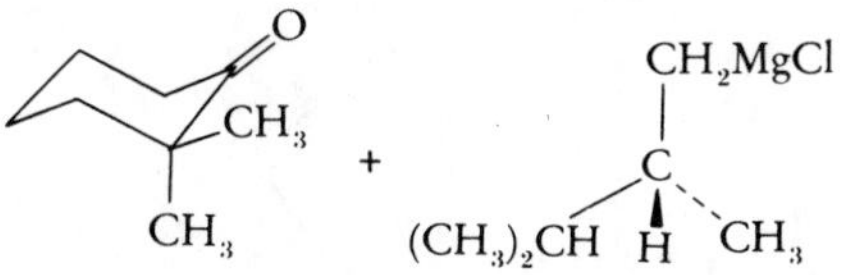

(c) $CH_3C-C(CH_3)_3 + S-CH_3CH_2CH(CH_3)CH_2CH_2MgCl$
$\quad\ \ \overset{\displaystyle\|}{O}$

3. Use Cram's rule or Prelog's rule to predict the major product in the following reactions.

(a) PhLi +

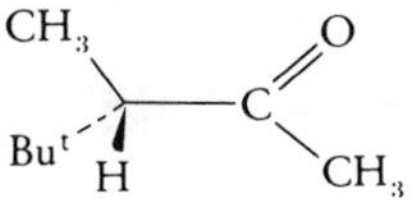

(b) $CH_3CH_2Li +$

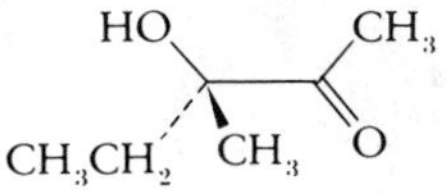

(c)

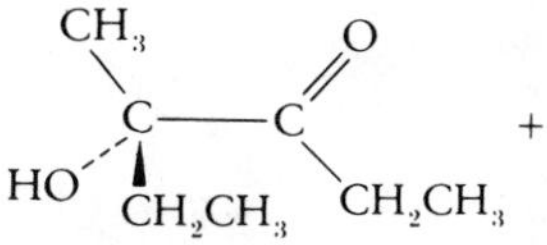

$+ CH_3Li$

(d) $S-PhCOCOOCH(CH_2CH_3)CPh_3 + CH_3CH_2MgBr$

Appendix

The Cahn–Ingold–Prelog sequence rules

A. Chirality of tetrahedral compounds

The R, S system was designed by R.S. Cahn, C. Ingold and V. Prelog as a method for specifying the configuration of chiral tetrahedral molecules, with a single chiral centre, in an unambiguous way. It has also been extended to other chiral systems and some diastereoisomeric systems.

Chiral molecules with a single chiral centre will be dealt with first. The rules depend on the assignment of *priorities* to the individual ligands attached to the chiral centre. The ordering of the ligands gives rise to a *descriptor*, R or S which prefixes the name of the compound. From this descriptor stereo-diagrams can be produced to illustrate the configuration of the compound.

The rules giving rise to the descriptors are:

(1) The ligands are assigned a priorities and numbered from one to four so that the *decreasing* priority is,

$$1 > 2 > 3 > 4$$

(2) The ligands are viewed from the side *opposite* the lowest priority ligand, 4, and the direction of decreasing priority determined. If the direction of decreasing procedure in viewing from 1–3 is clockwise, the configuration is designated R (rectus), and conversely an anticlockwise ordering is designated S (sinister). This is illustrated in Figure A1.1.

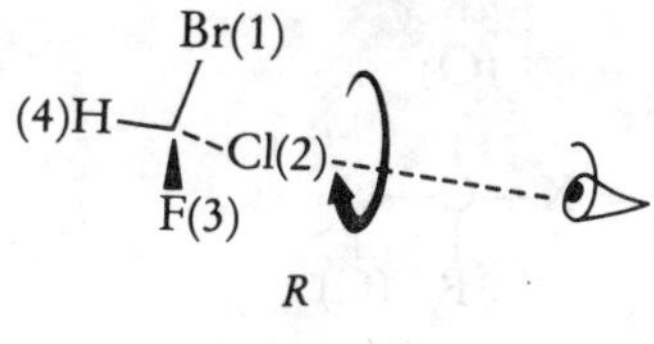

Figure A1.1.

213

The major rules for the determination of precedence are:

1. The atoms of higher *atomic number* have higher precedence (e.g. $F > O > N > C > B$).

2. For atoms of the same atomic number but different *atomic mass*, the higher mass number takes precedence (e.g. $T > D > H$).

3. Atoms attached directly to the central atom *must* be sequenced first. If two atoms of identical priority are attached to the central atom then each of the atoms attached to that must be compared. First look for the atoms of *highest* priority, e.g.

$$\underset{\underset{H}{|}}{\overset{\overset{H}{|}}{X-C-Cl}} > \underset{\underset{H}{|}}{\overset{\overset{H}{|}}{X-C-H}} \text{ where X is the central, chiral atom.}$$

If the highest priority atoms are identical, compare the next highest priority atoms, e.g.

$$\underset{\underset{F}{|}}{\overset{\overset{H}{|}}{X-C-Cl}} > \underset{\underset{H}{|}}{\overset{\overset{H}{|}}{X-C-Cl}}.$$

If no difference is encountered at the first series of atoms, then the process is followed, along the chains of highest priority, until a difference is found,

$$\text{e.g. } \underset{\underset{CH_3}{|}}{X-CH_2-CH-CH_2Cl} > \underset{\underset{CH_3}{|}}{X-CH_2-CH-CH_3}$$

When the central, chiral atom is part of a ring system, the sequence rules are exactly the same. Each branch of the ring is followed until a difference can be detected, e.g.

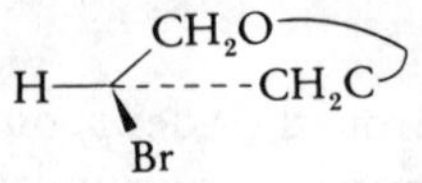

$$H-\underset{\underset{Br}{\Large\blacktriangledown}}{\overset{CH_2O}{\underset{-----CH_2C}{<}}}$$

$$R$$

4. Multiply bonded atoms are always treated as if they were *four* coordinate by adding replica atoms of the same type,

$$\text{e.g. } \underset{\underset{R}{|}}{X-C=O} \equiv \underset{\underset{R}{|}}{\overset{\overset{(O)}{|}}{X-C-O}} \quad (C)$$

214

The atoms in brackets are the replicas and are treated as bare atoms, surrounded by three ligands of *zero* priority,

$$\text{e.g.} \quad X-\underset{\underset{\text{CH}_3}{|}}{\overset{\overset{\text{CH}_3}{|}}{C}}-\text{CH}_2 > X-\underset{\underset{\text{H}}{|}}{\overset{\text{H}}{C}}=\text{CH}_2 \quad (\equiv X-\underset{\underset{\text{H}}{|}}{\overset{\overset{(\text{C})}{|}}{C}}-\text{CH}_2)$$

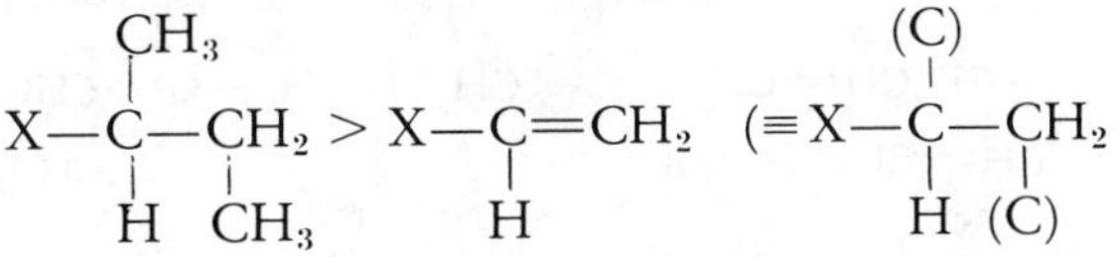

Table A1.1 gives a list of common atoms and groups in order of increasing priority.

Table A1.1. Atoms and groups of increasing priority according to the Sequence rules

1 —H	21 —COCH$_3$
2 —CH$_3$	22 —COC$_6$H$_5$
3 —CH$_2$CH$_3$	23 —COOH
4 —CH$_2$CH$_2$CH$_3$	24 —NH$_2$
5 —CH$_2$CH$_2$CH$_2$CH$_3$	25 —$\overset{+}{\text{N}}$H$_3$
6 —CH$_2$CH$_2$CH$_2$CH$_2$CH$_3$	26 —NHCH$_3$
7 —CH$_2$CH$_2$CH(CH$_3$)$_2$	27 —NHC$_2$H$_5$
8 —CH$_2$CH(CH$_3$)$_2$	28 —N(CH$_3$)$_2$
9 —CH$_2$—CH=CH$_2$	29 —NO
10 —CH$_2$—C≡CH	30 —NO$_2$
11 —CH$_2$C$_6$H$_5$	31 —OH
12 —CH=CH$_2$	32 —OCH$_3$
13 —C$_6$H$_{11}$	33 —OCH$_2$C$_6$H$_5$
14 —CH=CHCH$_3$	34 —OC$_6$H$_5$
15 —C(CH$_3$)$_3$	35 —F
16 —C≡CH	36 —SH
17 —C$_6$H$_5$	37 —S(O)CH$_3$
18 —C≡CH	38 —S(O)$_2$CH$_3$
19 —C$_6$H$_4$CH$_3$—o	39 —Cl
20 —CHO	40 —Br
	41 —I

Using the rules, the following priorities may be assigned,

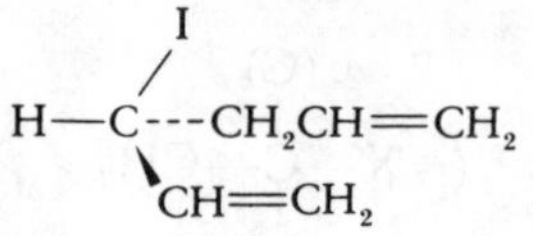

$$
\underset{S}{\overset{I}{H-\overset{|}{\underset{CH=CH_2}{C}}---CH_2CH=CH_2}}
\qquad
\underset{S}{\overset{C(CH_3)_3}{CH_3CH_2CH_2-\overset{|}{\underset{N(CH_3)_2}{C}}---CHO}}
$$

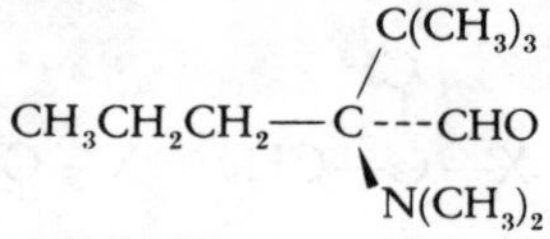

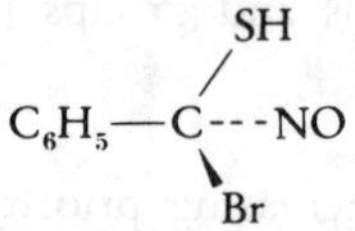

$$
\underset{R}{\overset{SH}{C_6H_5-\overset{|}{\underset{Br}{C}}---NO}}
$$

B. Extension of the rules to compounds with more than one chiral centre

The rules for compounds with two or more chiral centres are identical to those above; but *each* chiral centre must be sequenced separately.

One more priority rule is necessary and it states that,

$R > S$

Some examples are,

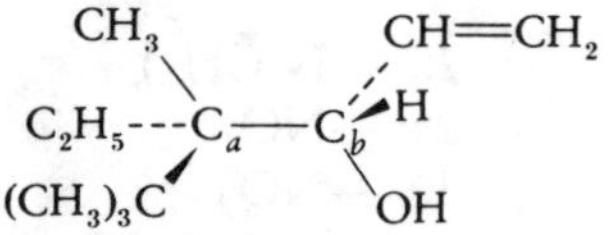

$$
\underset{S}{\overset{CH_3}{\underset{(CH_3)_3C}{C_2H_5---\overset{|}{C_a}-\overset{CH=CH_2}{\underset{OH}{C_b{\diagup}^H}}}}}
\qquad\qquad
\overset{HC=CH_2}{\underset{}{|}}
$$

Where, for C_a the priorities are $CH_3 < C_2H_5 < C(CH_3)_3 < -CHOH$; and for C_b $H < -CH=CH_2 < -C(CH_3)(C_2H_5)(C(CH_3)_3) < -OH$.

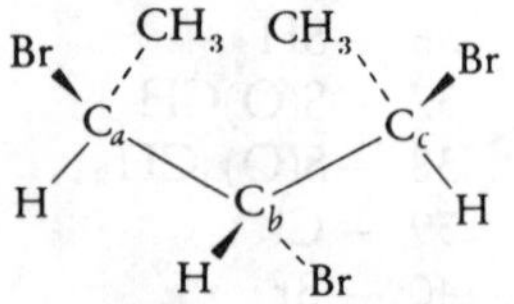

$$
\underset{S\quad\quad R\quad\quad R}{\overset{CH_3\ \ CH_3}{\overset{Br\diagdown\ |\quad|\ \diagup Br}{\underset{H}{\overset{C_a}{\diagup}}\ \underset{H\quad Br}{\overset{C_b}{}}\ \underset{}{\overset{C_c}{\diagdown}}\ H}}}
$$

The central carbon atom is designated R as $C_c > C_a$. The diastereoisomeric compound,

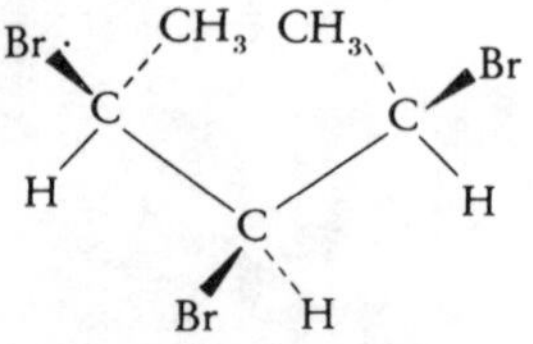

$$
\underset{S\quad\quad S\quad\quad R}{\overset{CH_3\ \ CH_3}{\overset{Br\diagdown\ |\quad|\ \diagup Br}{\underset{H}{\overset{C}{\diagup}}\ \underset{Br\quad H}{\overset{C}{}}\ \underset{}{\overset{C}{\diagdown}}\ H}}}
$$

216

has the *S,S,R* configuration. Both *S,R,R*-2,3,4-tribromopentane and *S,S,R*-2,3,4-tribromopentane are *meso*-compounds, with mirror planes of symmetry passing through the 3-carbon atom and its H and Br atoms.

C. Molecules with axial chirality

The same sequence rules are used for axially chiral molecules but the method of viewing is different. The structure in Figure A1.2 shows the extended tetrahedron on which axial chirality is based. The axis, arbitrarily marked as the *z* axis, is the reference axis, for viewing, when assigning a configuration. The ligands are assigned priorities with the additional rule that ligands *nearer* the eye *always* take priority over the further ligands. Now the descriptor *R* or *S* can be determined by following the sequence 1–3 in the usual way. *It does not matter from which end of the molecule the view is taken.*

Figure A1.2.

For example,

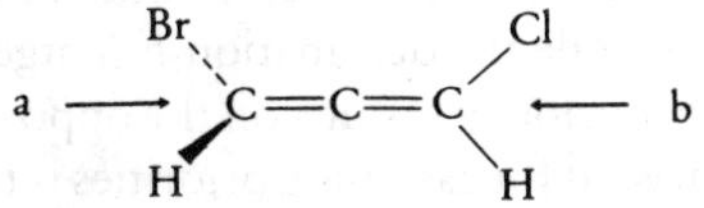

$$Br\overset{\cdot}{\underset{H}{\diagdown}}C=C=C\overset{Cl}{\underset{H}{\diagup}}$$

a ⟶   ⟵ b

can be viewed in two ways,

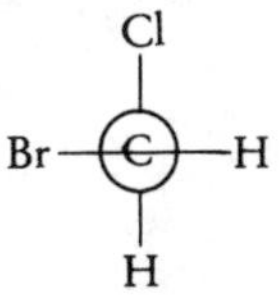

$$Br-\overset{Cl}{\underset{H}{\ominus}}-H \quad \text{from direction a,}$$

which on assignment of priorities gives,

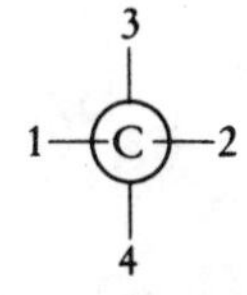

$$1-\overset{3}{\underset{4}{\odot}}-2$$

S

The allene is therefore designated *S*, by this method; now we can check that the configuration is the same when viewed along direction b,

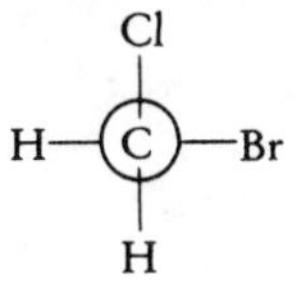

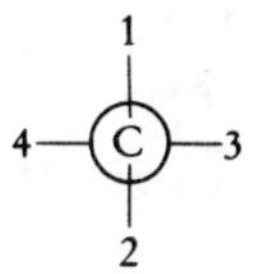

$$H-\overset{Cl}{\underset{H}{\odot}}-Br \qquad 4-\overset{1}{\underset{2}{\odot}}-3$$

S

217

This confirms that the modified system is unambiguous in its assignment of configuration to molecules with axial chirality.

D. The description of diastereoisomers arising from restricted rotation about π-bonds

The system described below is used for, alkenes, imines, oximes, amides, etc.

The general system can be written,

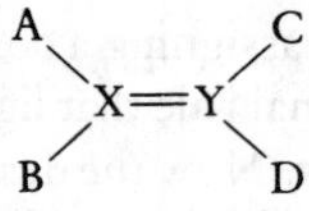

$$
\begin{array}{ccc}
A & & C \\
& \diagdown \quad \diagup & \\
& X=Y & \\
& \diagup \quad \diagdown & \\
B & & D
\end{array}
$$

The order of priorities of the groups on X and Y are determined separately, to give, say,

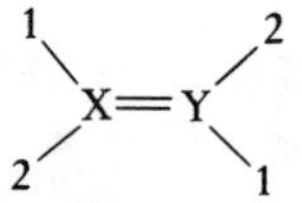

$$
\begin{array}{ccc}
1 & & 2 \\
& \diagdown \quad \diagup & \\
& X=Y & \\
& \diagup \quad \diagdown & \\
2 & & 1
\end{array}
$$

Now the relative dispositions of the two higher priority ligands (1) is compared. If they are on the *same* side of the double bond the diastereoisomer is designated Z (zusamman)—while the designation E (entgegen) is given to the diastereoisomer with higher priority groups on the opposite sides.

The only extra rule that is needed for assigning priorities is that lone pairs of electrons have a lower priority than H.

Some examples are,

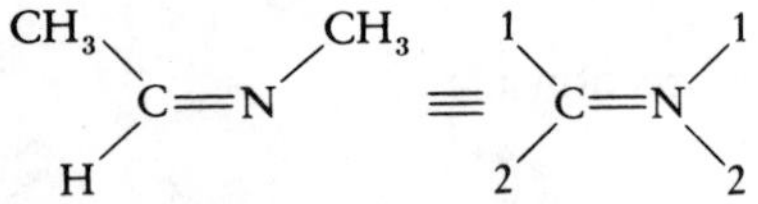

$$
\begin{array}{ccccc}
CH_3 & & CH_3 & 1 & & 1 \\
\diagdown & & \diagup & & \diagdown & \diagup \\
& C=N & & \equiv & C=N & \\
\diagup & & & & \diagup & \diagdown \\
H & & & 2 & & 2
\end{array}
$$

This diastereoisomer is designated Z and,

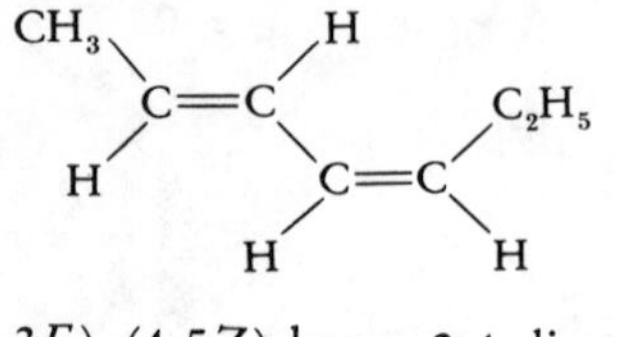

$$
\begin{array}{cccc}
CH_3 & & H & \\
\diagdown & & \diagup & \\
& C=C & & \\
\diagup & & \diagdown & \\
H & & C=C & C_2H_5 \\
& & \diagup \quad \diagdown & \\
& & H \qquad H &
\end{array}
$$

is called (2,3E), (4,5Z)-hepta-2,4-diene.

Answers to problems

Answers to problems

Chapter 1

2.

Central atom	Coordination number	Name of geometry	Idealized bond angles
C	4	tetrahedral	109.5°
C	3	trigonal	120°
C	2	digonal	180°
N	4	tetrahedral	109.5°
N	3	pyramidal	109.5°
N	2	trigonal	120°
O	2	bent	109.5°

5. (a) The angles are almost the ideal values of 120°. The small deviation is probably due to the slight steric interaction of the CH_2 groups with the neighbouring hydrogen atoms.

(b) As with CH_2Cl_2 the ClCCl bond angle is small. The electron withdrawing properties of the Cl atom reduces electron–electron repulsions between neighbouring C—Cl bonds and allows the Cl—C—C bonds to widen to reduce steric strain.

(c) The C—O—H bond in methanol is reduced below 109.5° primarily as a result of the electron–electron repulsions between the two lone pairs. This is a manifestation of the Thorpe–Ingold effect.

Chapter 2

2. The sp^3 hybrid orbitals give bond angles of 109° 28', which are closer to the angles of 107.3° and 104.5° found in ammonia and water than sp^2 (120°) or no hybridization (90°). The small closing of these bond angles is accounted for by repulsion of the lone pairs in sp^3 orbitals.

3. Even if you manage to construct the skeleton of 2.13(a) without breaking or bending the straws you will find that the p orbitals that should overlap to form the π-bond are at 90°. In other words there is no overlap and no double bond.

4. Again, amide delocalization depends on the coplanarity of the nitrogen lone pair orbital (sp^3 before overlap—p after overlap in a 'normal' amide) and the orbitals forming the carbonyl π-bond. A model of the bridgehead nitrogen compound illustrates that overlap would not be possible between the nitrogen and carbonyl p orbitals.

Chapter 3

1.

2. $IC \equiv CH$ has one C_x axis (infinite order)
$H{-}C \equiv C{-}I \text{------} C_x$
3.

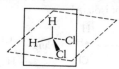

4. $H_2C{=}C{=}CH_2$ has a S_4 axis, running through the three carbon atoms. CH_4 also has S_4 axes.
5. Benzene and ethene are the only two molecules in the list with centres of symmetry.

Chapter 4

1. (a) diastereoisomers
   (b) constitutional isomers
   (c) homomers—(models if unsure)
   (d) enantiomers—(models)
   (e) enantiomers
   (f) homomers
   (g) enantiomers
2. (a) homomers
   (b) homomers
   (c) enantiomers
   Four different substituents are required for a tetrahedral molecule to be chiral.
3. (a) enantiomers (therefore chiral molecules)
   (b) diastereoisomers, only the second, *trans* structure is chiral
   (c) homomers
   (d) diastereoisomers, *both* molecules are chiral. The oxygen atom removes the mirror plane of symmetry in the *cis* isomer.

Chapter 5

1. **2** is another conformation of **1** but is at an energy maximum, as it is eclipsed. **2** is therefore not a conformer of **1**, as conformers are species at energy minima.
   3 is a conformer of **1**, a rotation about the central C—C bond enables the interconversion of **1** and **3**, and **3** is staggered and at an energy minimum.
   4 is similarly a conformer of **1** (and **3**).
   5 is not a conformer of **1**. No amount of rotation about any bonds can produce a relationship closer than diastereoisomeric. **5** and **1** are diastereoisomeric conformations. The compounds represented by **5** and **1** are also diastereoisomeric.
2. The pair of structures in (a) are diastereoisomeric but rotation about the central bond enables the two structures to become identical. The two compounds are therefore *homomers*.
   The pair of structures in (b) is enantiomeric but rotation about the central bond can produce homomeric structures. The compounds represented by the structures in (b) are therefore homomers and the structures themselves represent conformational enantiomers.
   The pair of structures in (c) are diastereoisomeric but they can be made enantiomeric by rotation. The compounds are therefore enantiomers, but the structures are diastereoisomeric.

3. Conformers are freely interconvertible, not isolable species at energy minima. They can be observed but not isolated.

e.g. [structures with Cl, H, H and Cl, Cl, H] and

Stereoisomers, are compounds with identical constitutions but different spatial arrangement of ligands. Enantiomers and diastereoisomers are both sub classes of stereoisomerism. Stereoisomers are not readily interconvertible and may be isolated as stable compounds.

e.g. enantiomers [structure H, Br, Cl, F] and [structure Br, H, Cl, F]

diastereoisomers [structure CH_3, CH_3, H, H] and [structure CH_3, H, H, CH_3]

Atropisomers are stereoisomers that can be isolated as pure substances but are formally interconvertible by rotations about single bonds. The structures in Figure 5.10 are examples of atropisomers.

Invertomers are enantiomeric structures that can usually be readily interconverted by non-bond-breaking inversions of configurations. Amines of formula RR'R''N exist as equimolar quantities of rapidly interconverting enantiomers.

Rotamers are diastereoisomeric structures that are freely interconvertible by rotations about partial double bonds, as in amides RCONR'R''.

Chapter 6

1. The staggered conformations of pentane (carbon skeletons only) are:

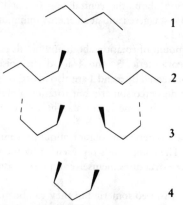

Conformation 4 is forbidden as two hydrogen atoms are competing for the same space.

Conformation **1** is the enthalpy favoured, all-antiperiplanar conformation, in which there are no *gauche* or *syn* interactions. The structures **2** are conformational enantiomers and therefore of equal energy. There is one synclinal bond in these conformations compared with two synclinal interactions in each of the conformers shown in **3**. The structures **3** are therefore the highest in enthalpy of the allowed conformations.

The interactions in butane are often used as comparisons for steric interactions in other compounds, where it is assumed that the order of decreasing interaction is,

synperiplanar > gauche (synclinal) > antiperiplanar

2. The Newman projections are:

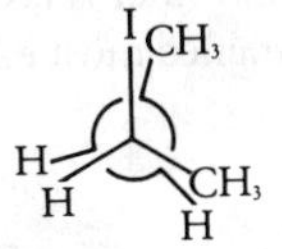

a 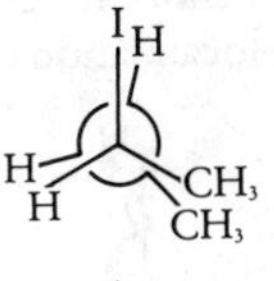   b

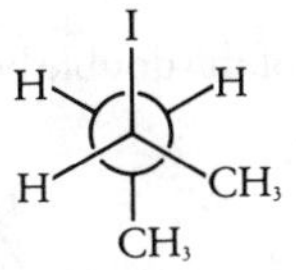

c 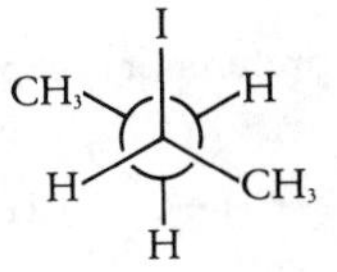   d

These are obtained by comparing van der Waals radii and assuming that $I > CH_3 > H$. The most severe steric interaction will occur when I is eclipsed by the CH_3 group.

The plot of energy against dihedral angle is unsymmetrical because the direction of rotation is important. Rotation about the 2,3 bond by 60° clockwise is energetically different from a 60° anticlockwise rotation, owing to the chirality of C-2. The *S* enantiomer of 2-iodobutane would show a mirrored curve.

3. (a)

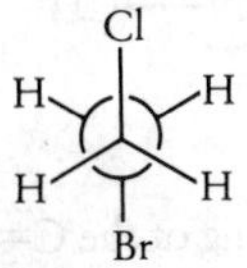

Both dipolar repulsions and steric effects should favour the antiperiplanar conformation.

3. (b)

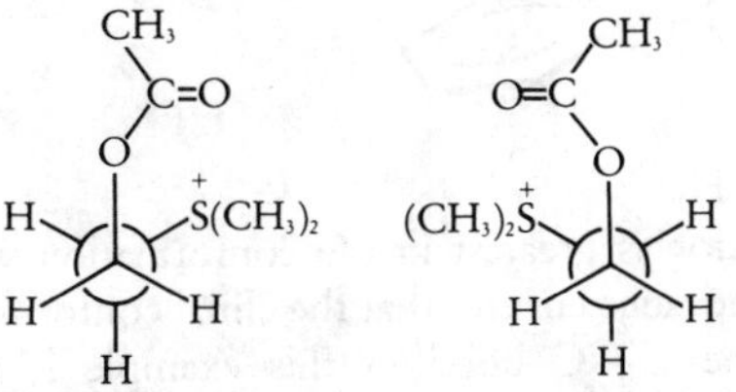

Electrostatic interaction between the positively charged —$\overset{+}{S}Me_2$ group and the partially negatively charged oxygen atoms stabilizes the synclinal conformation.

223

3. (c)

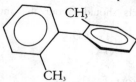

In 2-fluoroethanol the synperiplanar conformations are stabilized by intra-molecular hydrogen bonding.

3. (d)

The 2,3-s-*trans*,4,5-s-*trans* conformer is favoured for steric reasons. Planarity and hence delocalization can be attained most easily from this conformation.

3. (e)

This is preferential eclipsing of the double bond.

3. (f)

$$CH_3-C \overset{O}{\underset{O}{\diagdown}} CH_3$$

Once again the carbonyl C=O bond is eclipsed by a methyl group.

3. (g)

$$H-C \overset{O}{\diagdown} CH_2 \\ F$$

The dipole–dipole repulsion in the eclipsed conformer are strong enough to destabilize that conformation with respect to the *anti* conformation shown.

3. (h)

$$H-C \overset{O}{\diagdown} N-CH_3 \\ CH_3$$

Conjugation is maximized with planar nitrogen and eclipsing of the C=O bond.

3. (i)

Although conjugation is greatest in the conformation with both benzene rings coplanar, steric hindrance ensures that the clinal conformation has lowest energy. Rotation about the C—C bond in this example is restricted, but separate enantiomers cannot be isolated.

Chapter 7

1. There are two sets of 1,3-dichlorocyclohexanes—*cis* and *trans*. The *cis* isomer exists in two conformations,

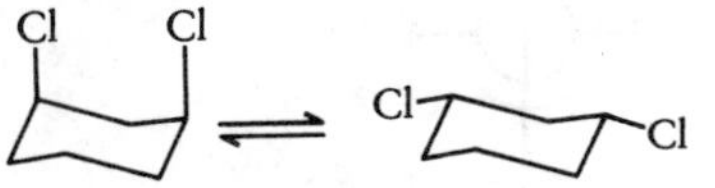

Both the a,a and e,e conformers of *cis*-1,3-dichlorocyclohexane have mirror planes of symmetry (running through the 2 and 5 carbon atoms). This diastereoisomer is achiral.

The *trans* isomer exists in two enantiomeric configurations, each one can undergo a ring inversion to give an indistinguishable conformer, but 2 and 3 are not inter-convertible by ring flips. (models)

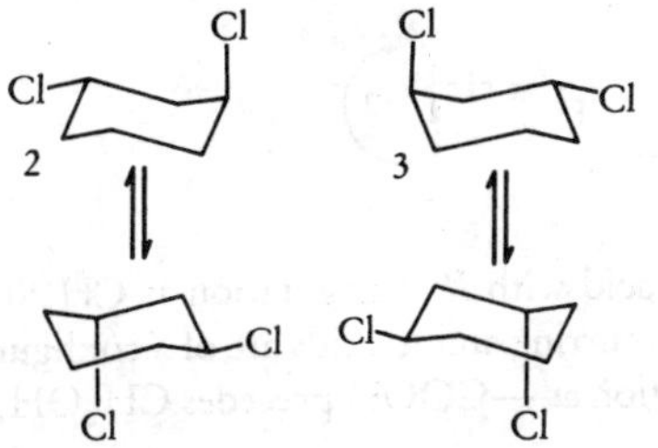

There are therefore 3 separate stereoisomers of 1,3-dichlorocyclohexane; the *meso cis* isomer and two enantiomeric *trans*-1,3 isomers.

2. There are two isomers of 1,4-C_6H_{10}(OH)COOH, *cis* and *trans*. The hydroxy and carboxyl groups in both the a,a and e,e *trans* conformations are too remote to interact. Models show that the groups can interact in the *cis-boat* form, to give

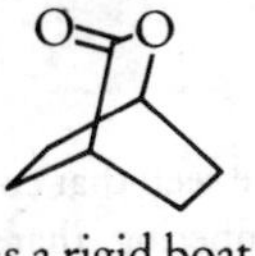

which has a rigid boat conformation for the cyclohexane ring.

3.

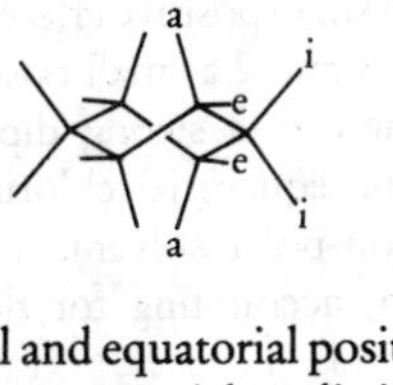

The axial and equatorial positions are sometimes labelled *tax,* and *teq* for twist-axial and twist-equatorial to distinguish them from the chair axial and equatorial positions.

4. There are two, non-degenerate conformations of —N—H,

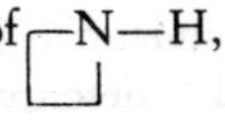

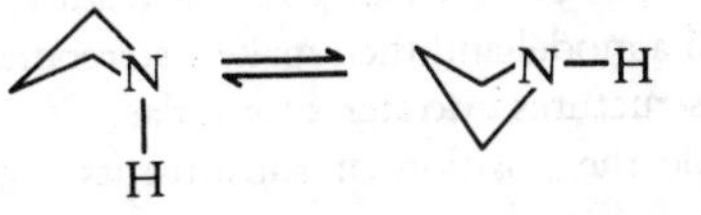

The equatorially substituted NH conformation is favoured by about 1 kJ mol^{-1}.

Chapter 8

1. (a) Axis of chirality.

(b) Plane of chirality. The sequence rules for planes of chirality are not covered in the Appendix.

(c) Centre of chirality.

This is an example of an L amino acid with R configuration as CH_2SH precedes —COOH. Most other naturally occurring amino acids are of S configuration.

(d) Centre of chirality. S configuration as —COOH precedes CH_2OH.

(e) Axis of chirality, R (cf 1a).

3. Two conformations for 2R,5R-*trans*-2-chloro-5-methylcyclohexanone can be drawn,

Without a complete analysis of the Cotton effect, it can be seen that conformation 1, has an axial C-6 substituent, according to the usual numbering, that is predicted to have a strongly negative effect. The equatorial chlorine in 2 will have little effect and the 5-equatorial methyl, $-x, -y, +z$, will have a small positive effect. Therefore 1 is predicted to have a large negative Cotton effect and 2 a small positive effect. Although 1 is disfavoured on steric grounds there is a strong dipole–dipole repulsion that destabilizes 2. In polar solvents the equatorial chlorine and the carbonyl groups are shielded by solvation. In non-polar solvents some of the molecules are forced into the axial configuration, accounting for the negative Cotton effect in octane.

4. A brief summary of model use is,

chirality Model and mirror image model can be compared for superimposability more easily than formulae may be compared.

configuration R and S descriptors may be translated into a readily assimilated structure. It is often easiest to build a model and then make a stereodrawing from the model when translating named structures into stereoformulae.

Cotton effect. A model will enable the position of substituents to be located quickly.

Chapter 9

2. (a) The right and left-handed helices of polyisotactic propylene are *enantiomeric*, as the chains are achiral. The Cotton effect for each helix should be equal and opposite.

(b) The two helices of poly(L-alanine) are diastereoisomeric, owing to the chirality of the constituent amino acids. The two Cotton effects would therefore be different, but it is not possible to say, *a priori*, whether they would be of opposite signs.

It would not be possible to observe a Cotton effect for polyisotactic propylene, as there are equal numbers of the degenerate, right and left-handed helices, and they are also in dynamic equilibrium.

A Cotton effect would be observed, under any circumstances for poly(L-alanine) as there is an asymmetrically perturbed chromophore, even in a random sample.

3. There are three basic structures for poly $(CH_3CH=CD)$

(a) achiral, (*threo*), diisotactic,

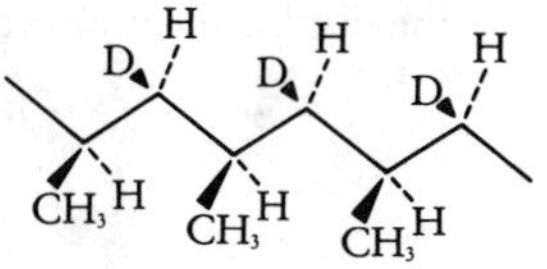

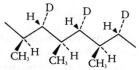

(b) achiral (*erythro*) diisotactic (models show that *long* chains will be effectively achiral, neglecting end groups).

(c) achiral, disyndiotactic.

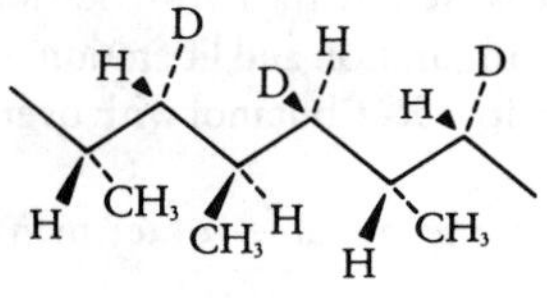

Chapter 11

1. (a) As **1** and **2** both react at unchanged rates in ethanol and in the presence of alkoxide ion an S_N1 reaction is proposed.

(b) A planar carbonium ion is the intermediate in S_N1 solvolysis. The carbonium ions from both **1** and **2** would be strained, as models will show, but the extra carbon atom in the bridge of **1** allows planar intermediate to be formed with less strain. The activation energy is lower for **1**, and hence the rate of solvolysis of **1** is greater.

(c) Inversion of configuration of **1** and **2** is not possible on steric grounds. The incoming nucleophile would have to get inside the cage to attack the C—Cl carbon from the rear.

2. The two possible conformations of *trans* 1,3-BrC$_6$H$_{10}$COO$^-$ are,

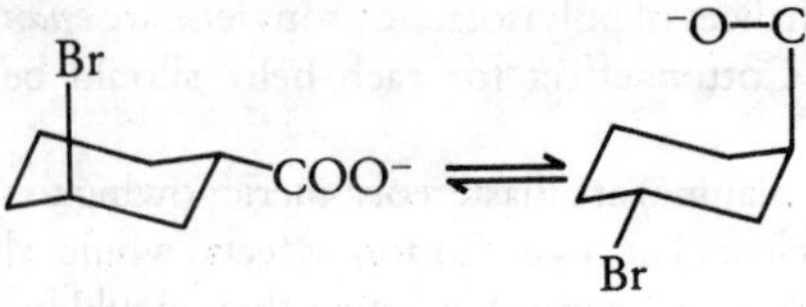

Conformation 4 has the carboxylate anion suitably arranged for S_N2 displacement of bromide ion to give the lactone 5,

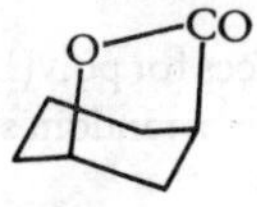

The two *cis* conformers 6 and 7 cannot allow the displacement of bromide ion as frontside attack would be necessary,

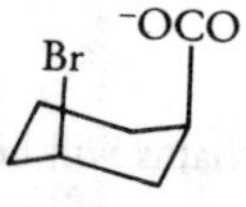

3. *R*-2-chlorobutane reacts slowly with hydroxide ion in a bimolecular substitution to give the product alcohol with inversion of configuration.

Under iodide ion catalysis conditions, the reaction takes place in two steps. First the iodide ion displaces chloride with inversion of configuration, to give *S*-2-iodobutane. This intermediate product is attacked by the hydroxide ion to give the product alcohol, again with inversion of configuration and liberation of iodide ion. The product with iodide ion present is therefore *R*-2-butanol with overall *retention* of configuration.

4. The following reaction schemes represent one possible route; many others are possible.

(a) $S-CH_3(C_3H_7)CHOTs+Cl^- \xrightarrow[\text{solvent}]{\text{non-polar}} R-CH_3(C_3H_7)CHCl$

(b) $R-CH_3(C_3H_7)CHOH+SOCl_2 \xrightarrow[\text{solvent}]{\text{non-ionizing}} R-CH_3(C_3H_7)CHCl$

(c) R or $S-CH_3(C_3H_7)CHCl+TSO^- \xrightarrow[\text{solvent}]{\text{ionizing}} R+S-CH_3(C_3H_7)CHOTs$

(d) $S-CH_3(C_3H_7)CHCl+TSO^- \xrightarrow[\text{solvent}]{\text{non-ionizing}} R-CH_3(C_3H_7)CHOTs$

Examples of ionizing solvents are, DMF, DMSO, CH$_3$CN, and (CH$_3$)$_2$CO/

Chapter 12

1. (a) *syn*-elimination, Saytzeff product preferred.

(b) *E*1 elimination, non-stereoselective, but thermodynamically favoured product **1** will be in excess.

(c) *E*2 elimination, stereospecific, only anti-elimination possible. **2** is the Saytzeff product.

2. *E*2, anti-elimination cannot take place from **13**, in the preferred conformation shown (models), however ring flipping gives an *anti*-diaxial arrangement, so that *E*2 elimination can occur. Only one product is possible.

There are two *anti*-diaxial arrangements in **14**. Saytzeff orientation is preferred in *E*2 eliminations so **15** will predominate.

3. (a) Good nucleophile, weak base, secondary substrate S_N2 favoured.
   $\therefore$ (CH$_3$)$_2$CHCH(CN)CH$_3$.

(b) Strong base, good nucleophile, secondary substrate S_N2 and $E2$ favoured, $\therefore$ mixture of products. For elimination Saytzeff preferred, $(CH_3)_2C = CHCH_3$. Substitution product $(CH_3)_2CHCH(OC_2H_5)CH_3$.

(c) Strong base, weak nucleophile, secondary substrate, neutral leaving group. $E2$ elimination preferred, Saytzeff product $(CH_3)_2C = CHCH_3$.

(d) Strong base, moderate nucleophile, charged substrate, high temperature. Elimination preferred, Hofmann orientation. $(CH_3)_2CHCH = CH_2$ major product.

4. (a)

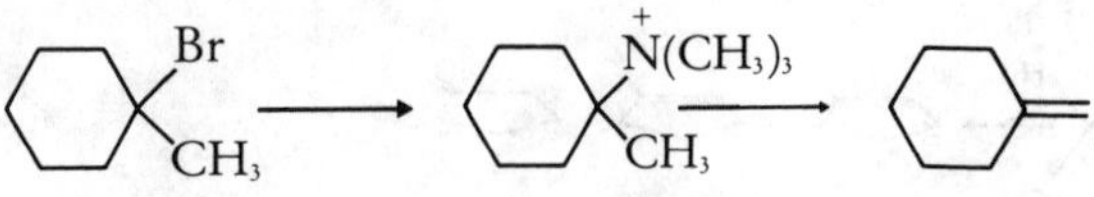

or

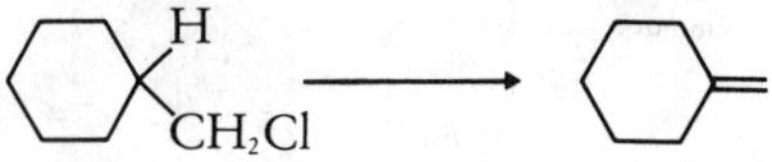

(b)

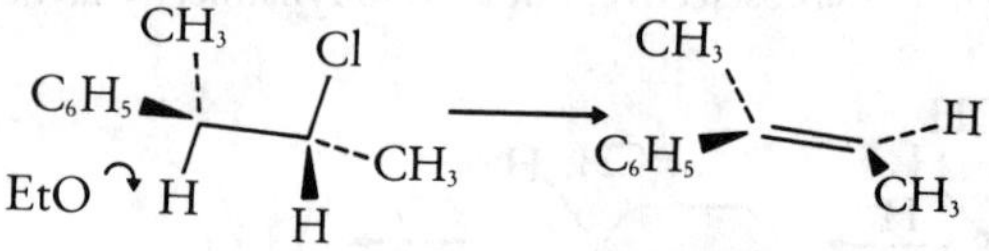

(c)

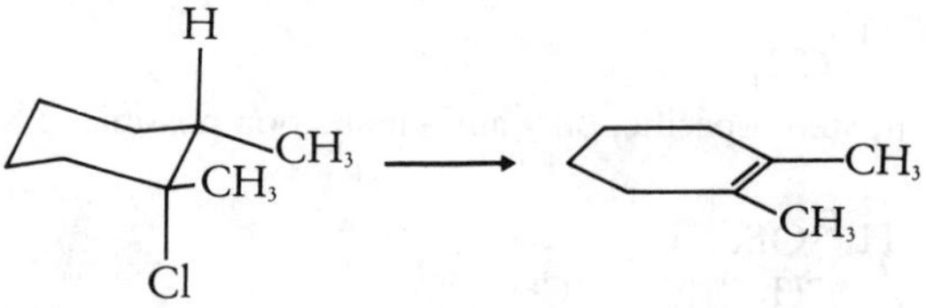

5. (a)

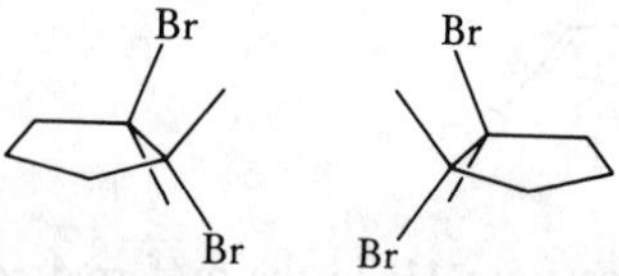

Anti-addition of Br_2. Enantiomers formed in equimolar quantities.

(b)

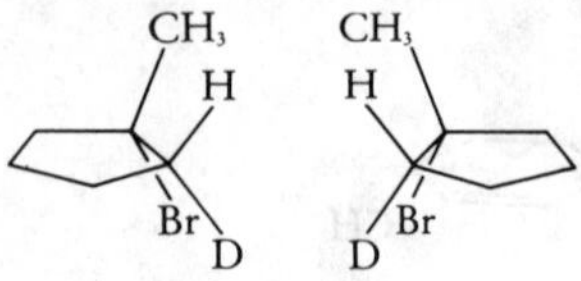

Regiospecific *anti*-addition of HBr—Markownikov's rule. Equimolar amounts of enantiomers.

(c)

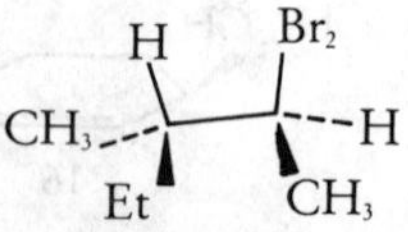

$+$ *enantiomer*

Regiospecific *syn*-addition. Equimolar amounts of enantiomers.

230

(d)

Chiral substrate–diastereoisomers formed by *anti*-addition.

Chapter 13

1. The electrocyclic ring opening of cyclobutenes is a four-electron reaction that is allowed thermally in the conrotatory mode. Two products could be produced by conrotatory opening,

E,E—hexa-2,4-diene *or* Z,Z-hexa-2,4-diene.
The Z,Z-diastereoisomer cannot adopt an s-Z conformation for steric reasons and the transition state leading to its formation is very highly hindered. The only product formed on thermolysis of *trans*-3,4-dimethylcyclobutane is *E,E*-hexa-2,4-diene.

2. As no information was given in the question about the nature of the reactions, thermal or photochemical both need to be examined. Electrocyclic ring opening is the most likely first step. Thermal conrotatory ring opening gives a hexatriene that is suitably arranged for an electrocyclic ring closure.

This alkene will ring close thermally in a disrotatory mode as the transition state is a six-electron system. Such a thermal, disrotatory closure gives the required product with a *trans* ring junction. One answer to the problem is therefore a thermal conrotatory ring opening followed by a thermal disrotatory ring closure.
Photochemical processes have opposite selection rules to thermal processes, it might appear that the same transformation could occur by a photochemical disrotatory ring opening followed by a photochemical conrotatory ring closure. Let us examine this sequence.
Although there are two theoretically allowed products of disrotatory ring opening of the substrate, by analogy with the first problem in this chapter, only one can be formed for steric reasons.

The *E*-stereochemistry about both of the newly formed double bonds precludes a further electrocyclic ring closure, either photochemical or thermal, as the termini of the π-system cannot get sufficiently close for reaction to take place.

3. The basis set for any 1,3 dipole is 3 p orbitals containing four electrons (as also found in amides for example). So the basis set for suprafacial–suprafacial cyclo-addition is,

This is a six-electron transition state with an even number of phase dislocations; it is therefore thermally allowed. The products of diazomethane addition to Z-but-2-ene are therefore the compound shown and its enantiomer.

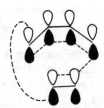

4. The basis set for a $_2\pi_a + _4\pi_a$ cycloaddition is,

No attempt has been made to describe the steric deformations necessary for such an addition but the required phase relationships are given. The transition state shown is a six-electron Hückel system with an even number of phase dislocations. It is therefore Hückel aromatic and thermally allowed.

5. The sigmatropic migration of a hydrogen atom in the compound shown involves a six-electron transition state and is therefore suprafacial (Table 13.3).

 Two conformations from which hydrogen can migrate are possible, and each has a different stereochemical result,

Each of the products shown is formed enantiomerically pure, since an antarafacial migration would be required for enantiomer formation.

Chapter 14

1. (a)

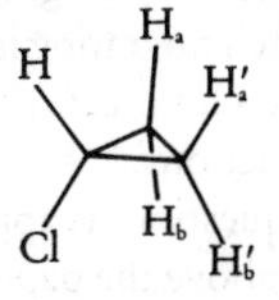

H_a and $H_{a'}$ are enantiotopic and so are H_b and $H_{b'}$. If H is substituted by chlorine then H_a and $H_{a'}$ produce enantiomers on substitution but H_b and $H_{b'}$ produce homomeric *cis*-1,2-dichlorocyclopropanes. It is necessary to ensure that different groups are used on application of the substitution rule, or extra symmetry may unwittingly be introduced.

(b) This is quite a complex example.

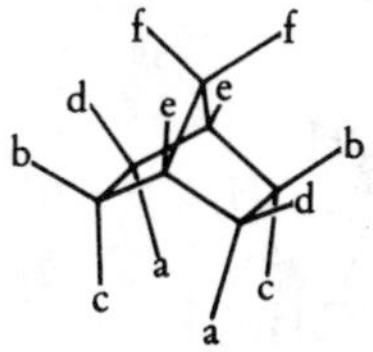

Each hydrogen atom is labelled with a letter only for simplicity. The hydrogen atoms with the same subscript are homotopic. In addition H_b and H_d are enantiotopic and H_a and H_c are also enantiotopic.

(c) The two faces of the carbonyl group are enantiotopic, as is demonstrated by reduction.

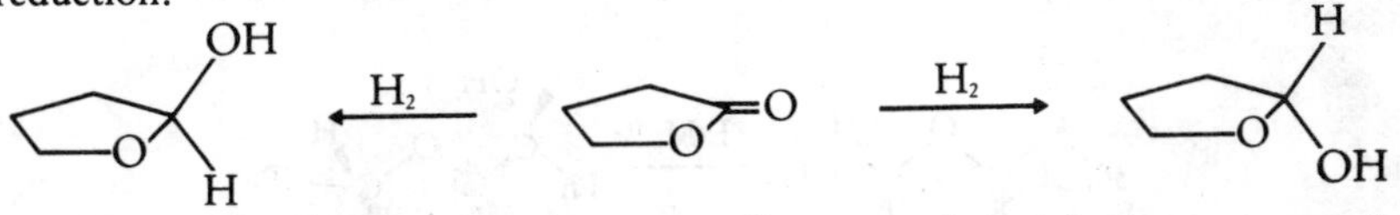

(d) The two faces of the carbonyl group are diastereotopic and there are two sets of diastereotopic hydrogen atoms. One set is the hydrogen atoms *cis* to the chiral group and the other is the *trans* hydrogen atoms.

2. (a) Enantioface differentiation. The favoured transition state and product are,

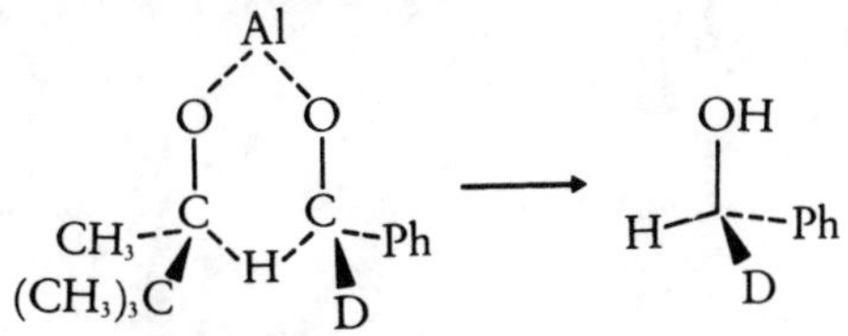

(b) Enantioface differentiation. The favoured transition state and product are,

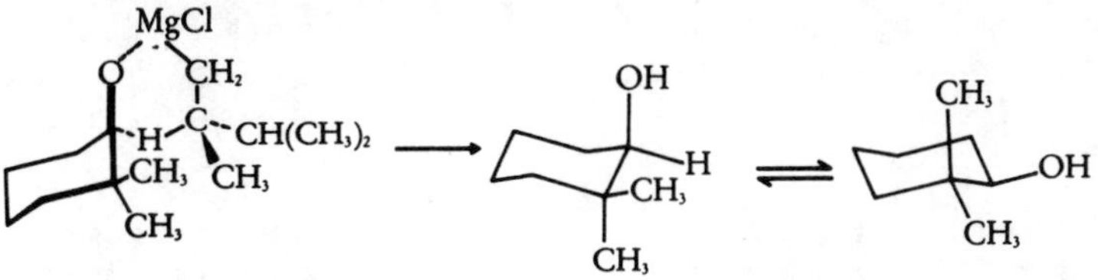

(c) There is no differentiation possible in this reaction. A drawing of the transition state shows that there is a choice of hydrogen atoms that can migrate, and each would give rise to one enantiomer. The probability of each migrating is identical so a racemic product is obtained. This illustrates that the chiral centre must be an intimate part of the transition state for differentiation reactions.

3. (a) The conformation of the substrate, as shown in the question, is appropriate for the operation of Cram's rule, and attack is from the face above the paper plane. The major product is therefore,

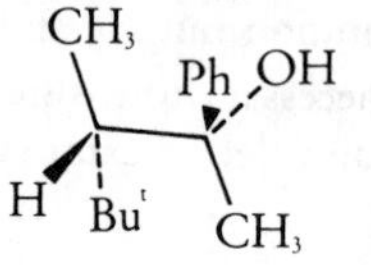

CH₃
Ph OH
H Buⁱ CH₃

(b) The modified Cram's rule applies here, as the OH group will complex with the metal and bridge the carbonyl.

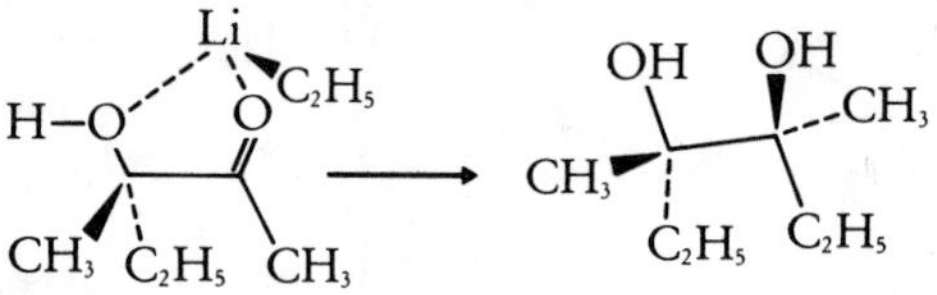

Li
H—O   C₂H₅
        O
CH₃ C₂H₅ CH₃

OH   OH
CH₃   CH₃
C₂H₅ C₂H₅

The enantiomer shown is the preferred product.

(c) Although superficially similar to the example in (b) the *meso* product is formed in this example.

(d) This is a typical example of the application of Prelog's rule.

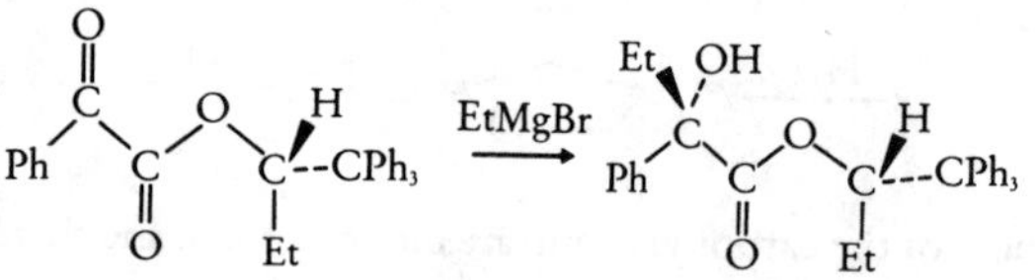

O
C
Ph   O   H
     C   C--CPh₃
     O   Et

EtMgBr

Et OH
C
Ph   O   H
     C   C--CPh₃
     O   Et

Some suggestions for
further reading

1   *Other stereochemistry texts of similar level:*

G. Natta and M. Farina, *Stereochemistry*, Longman, London, 1972.
J. Dale, *Stereochemistry and Conformational Analysis*, Verlag Chemie, *New York, 1978.*
G. Hallas, *Organic Stereochemistry*, McGraw-Hill, New York, 1966.
G. D. Gunstone, *Guidebook to Stereochemistry*, Longman, London, 1975.
E. L. Eliel, *Elements of Stereochemistry*, Wiley, New York, 1969.

2   *More advanced texts:*

E. L. Eliel, *Stereochemistry of Carbon Compounds*, McGraw-Hill, New York, 1962.
K. Mislow, *Introduction to Stereochemistry*, Benjamin, New York, 1966.
E. L. Eliel, N. L. Allinger, S.J. Angyal and G. A. Morrison, *Conformational Analysis*, Wiley, New York, 1965.
M. Nógrádi, *Stereochemistry, Basic Concepts and Applications*, Pergamon Press, Oxford, 1981.
D. H. R. Barton, *Principles of Conformational Analysis*, Angew Chem., 1970, 82, 827 (Review).

3   *Stereochemistry and reaction mechanisms:*

R. B. Woodward and R. Hoffmann, *The Conservation of Orbital Symmetry*, Verlag Chemie—Academic Press, 1970.
G. B. Gill and M. R. Willis, *Pericyclic Reactions*, Chapman and Hall, London, 1974.
R. W. Alder, R. Baker and J. M. Brown, *Mechanism in Organic Chemistry*, Wiley, New York, 1971.
C. K. Ingold, *Structure and Mechanism in Organic Chemistry*, 2nd Edition, Bell, London, 1970.
B. Breslow, *Organic Reaction Mechanisms: An Introduction*, Benjamin, New York, 1969.

4   *Organic structure determination:*

D. H. Williams and I. Fleming, *Spectroscopic Problems in Organic Chemistry*, McGraw-Hill, 1967.
J. E. Crooks, *The Spectrum in Chemistry*, Academic Press, 1978.

Index

237